Linked Data in Linguistics

Christian Chiarcos · Sebastian Nordhoff ·
Sebastian Hellmann
Editors

Linked Data
in Linguistics

Representing and Connecting
Language Data and Language Metadata

 Springer

Editors
Christian Chiarcos
Information Sciences Institute
University of Southern California
Marina del Rey, CA, USA

Sebastian Hellmann
Business Information Systems
University of Leipzig
Leipzig, Germany

Sebastian Nordhoff
Department of Linguistics
Max-Planck Institute for
Evolutionary Anthropology
Leipzig, Germany

ISBN 978-3-642-28248-5   e-ISBN 978-3-642-28249-2
DOI 10.1007/978-3-642-28249-2
Springer Heidelberg Dordrecht London New York

Library of Congress Control Number: 2012933118

Printed on acid-free paper

Springer is part of Springer Science+Business Media (www.springer.com)

Preface

The explosion of information technology has led to a substantial growth in quantity, diversity and complexity of web-accessible linguistic data. These resources become even more useful when linked. This volume provides an overview over recent developments, use cases, applications and recommendations for the application of the linked data paradigm to represent, exploit, store, and connect different types of linguistic data collections.

Recent relevant developments include: (1) Language archives for language documentation, with audio, video, and text transcripts from hundreds of (endangered) languages. (2) Typological databases with typological and geographical data about languages from all parts of the globe. (3) Development, distribution and application of lexical-semantic resources in Natural Language Processing. (4) Multi-layer annotations and semantic annotation of corpora by corpus linguists and computational linguists, often accompanied by the interlinking of corpora with lexical-semantic resources.

The general trend of providing data online is accompanied by newly developing possibilities to interconnect linguistic data and metadata. This includes general-purpose knowledge bases such as the DBpedia (a machine-readable edition of the Wikipedia), but also repositories with specific linguistic information about languages, as well as about linguistic categories and phenomena.

It is the challenge of our time to store, interlink and exploit this wealth of data, e.g., by modeling different language resources as Linked Data. The contributions assembled in this volume illustrate the band-width of applications of the Linked Data paradigm for representative types of language resources, including lexical-semantic resources, annotated corpora, typological databases as well as terminology and metadata repositories. The book includes representative applications from different fields, ranging from theoretical linguistics (e.g., typology) over applied linguistics (e.g., language documentation, lexicography, and corpus linguistics) to computational linguistics, Natural Language Processing and information technology.

This volume accompanies the Workshop on Linked Data in Linguistics (LDL-2012), held March 7th-9th, 2012 in Frankfurt a. M., Germany, organized by the

Open Linguistics Working Group (OWLG) of the Open Knowledge Foundation (OKFN). It assembles contributions of the workshop participants and, beyond this, it summarizes initial steps in the formation of a Linked Open Data cloud of linguistic resources, the Linguistic Linked Open Data cloud (LLOD).

Heidelberg and Leipzig,   *Christian Chiarcos*
December 2011   *Sebastian Nordhoff*
                                                                     Sebastian Hellmann

Acknowledgements

We would like to thank the organizers of the DGfS for supporting our work by integrating the workshop on Linked Data in Linguistics as a part of the 34th Annual Meeting of the German Linguistic Society (Deutsche Gesellschaft für Sprachwissenschaft, DGfS), March 7th-9th, 2012, in Frankfurt a. M., Germany. Further, we thank the Max-Planck Institute for Evolutionary Anthropology Leipzig and the LOD2 project – a large-scale integrating project co-funded by the European Commission within the FP7 Information and Communication Technologies Work Programme (Grant Agreement No. 257943) – at the Agile Knowledge Engineering and Semantic Web (AKSW) Lab of the University of Leipzig for their financial and organizational support.

We would like to express our gratitude to the participants of the Workshop on Linked Data in Linguistics, whose contributions are assembled in this volume, for sharing their ideas, insights and/or resources. Due to organizational reasons, a considerable number of papers could not be accepted, and we would also like to thank the authors of these contributions, hoping that they stay in contact with us and keep on working towards the application of the Linked Data paradigm in their research.

The choice of papers was not always easy, and judging only on the quality or innovativeness of the submissions, we would have liked to include more of them. The selection process did, however, not only rely on these two criteria, but also involved other goals, partially conflicting with each other, such as the representativeness for the respective community, maturity of the approach, and the overall goal to illustrate the diversity of recent approaches. We thank the program committee (a list is included after this preface) for their invaluable support and engagement in this process and their feedback on the different contributions.

We also thank our invited speakers, Nancy Ide and Martin Haspelmath, for presenting their work and sharing their experience from two of the primary prospective fields of applications of the Linked Data paradigm in Linguistics, i.e., Natural Language Processing and typological research. Additionally, Christian Kreutz introduced the Open Knowledge Foundation, i.e., the host organization of the Open Linguistics Working Group (OWLG) that organized the workshop. We also thank the members of the OWLG who indirectly contributed to the workshop and to this volume through discussions and the development of a common vision of a Linguistic Linked Open Data cloud.

Finally, we would like to thank Springer, for support and preparing the work.

Reviewing Committee

Anthony Aristar	Eastern Michigan University
Emily M. Bender	University of Washington
Hans-Jörg Bibiko	MPI-EVA Leipzig
Philipp Cimiano	CITEC, Universität Bielefeld
Alexis Dimitriadis	Universiteit Utrecht
Caroline Féry	Universität Frankfurt
Daniel Fleischhacker	Universität Mannheim
Jeff Good	University at Buffalo
Harald Hammarström	MPI-EVA Leipzig
Kees Hengeveld	Universiteit Amsterdam
Ernesto William de Luca	DAI-Lab, Technische Universität Berlin
Harald Lüngen	IDS Mannheim
Lutz Maicher	Fraunhofer MOEZ
John McCrae	CITEC, Universität Bielefeld
Gerard de Melo	MPI for Informatics, Saarbrücken
Pablo Mendes	FU Berlin
Steven Moran	University of Washington
Axel-C. Ngonga Ngomo	Universität Leipzig
Antonio Pareja-Lora	Universidad Complutense de Madrid
Cornelius Puschmann	Heinrich-Heine-Universität Düsseldorf
Felix Sasaki	DFKI Berlin, FH Potsdam
Stavros Skopeteas	Universität Bielefeld
Dennis Spohr	CITEC, Universität Bielefeld
Johanna Völker	Universität Mannheim
Menzo Windhouwer	MPI Nijmegen / Universiteit Amsterdam
Alena Witzlack-Makarevich	University of Zurich

Contents

Acronyms

API	Application programming interface
ASJP	Automated Similarity Judgment Program
ATLAS	Architecture and Tools for Linguistic Analysis Systems
CES	Corpus Encoding Standard
CLARIN	Common Language Resources and Technology Infrastructure
CMDI	CLARIN MetaData Infrastructure
DCR	Data Category Registry, see ISOcat
DGfS	Deutsche Gesellschaft für Sprachwissenschaft ('German Linguistic Society')
HTTP	Hypertext Transfer Protocol
ISO	International Organization for Standardization
ISOcat	Data Category Registry maintained by ISO TC37/SC4
GOLD	General Ontology for Linguistic Description
GrAF	Graph Annotation Format, XML linearization of the LAF
LAF	Linguistic Annotation Framework, upcoming standard developed by ISO TC37/SC4
LDL	Workshop on Linked Data in Linguistics (LDL-2012)
lemon	LExicon Model for ONtologies
LMF	Lexical Markup Framework, standard for NLP lexicons and machine-readable lexicons developed by ISO TC37/SC4
LOD	Linked Open Data
LLOD	Linguistic Linked Open Data cloud
MT	Machine Translation
NLP	Natural Language Processing
OKFN	Open Knowledge Foundation
OLAC	Open Language Archive Community
OWL	Web Ontology Language, RDF extension for ontologies
OWL/DL	OWL dialect to represent Description Logics
OWLG	Open Linguistics Working Group
PAULA	Potsdamer Austauschformat für Linguistische Annotationen ('Potsdam exchange format for linguistic annotations')

PID   Persistent Identifier
RDF   Resource Description Framework
RDFa   Resource Description Framework in attributes, a collection of attributes
             and processing rules for extending XHTML to support RDF
RDFS   RDF Schema
SGML   Standard Generalized Markup Language, predecessor of XML
SKOS   Simple Knowledge Organization Scheme, RDF extension for knowledge
             representation
SMT   Statistical Machine Translation
SPARQL   SPARQL Protocol and RDF Query language
SQL   Structured Query Language
TBX   TermBase eXchange, an XML format designed to allow the exchange of
             terminology databases between tools, developed on the basis of TMX
TEI   Text Encoding Initiative
TMX   Translation Memory eXchange, an XML format for the exchange of
             translation memory data
URI   Uniform Resource Identifier
URL   Uniform Resource Locator
XLIFF   XML Localization Interchange Format
XML   eXtensible Markup Language

Introduction and Overview

Christian Chiarcos, Sebastian Hellmann, and Sebastian Nordhoff

1 Linked Data in Linguistics

The explosion of information technology in the last two decades has led to a substantial growth in quantity, diversity and complexity of web-accessible linguistic data. These resources become even more useful when linked with each other, and the last few years have seen the emergence of numerous approaches in various disciplines concerned with linguistic resources.

It is the challenge of our time to store, interlink and exploit this wealth of data accumulated in more than half a century of computational linguistics (Dostert, 1955), of empirical, corpus-based study of language (Francis and Kucera, 1964), and of computational lexicography (Morris, 1969) in all its heterogeneity.

A crucial question involved here is the **interoperability** of the language resources, actively addressed by the community since the late 1980s (Text Encoding Initiative, 1990), but still a problem that is partially solved at best (Ide and Pustejovsky, 2010). A closely related challenge is **information integration**, i.e., how heterogeneous information from different sources can be retrieved and combined in an efficient way.

With the rise of the Semantic Web, new representation formalisms and novel technologies have become available, and, independently from each other, researchers in different communities have recognized the potential of these devel-

Christian Chiarcos
Information Sciences Institute, University of Southern California, 4676 Admiralty Way # 1001, Marina del Rey, CA 90292 e-mail: chiarcos@daad-alumni.de

Sebastian Hellmann
Universität Leipzig, Fakultät für Mathematik und Informatik, Abt. Betriebliche Informationssysteme, Johannisgasse 26, 04103 Leipzig, Germany e-mail: hellmann@informatik.uni-leipzig.de

Sebastian Nordhoff
Department of Linguistics, Max Planck Institute for Evolutionary Anthropology, Deutscher Platz 6, 04103 Leipzig, Germany e-mail: sebastian_nordhoff@eva.mpg.de

C. Chiarcos et al. (eds.), *Linked Data in Linguistics*,
DOI 10.1007/978-3-642-28249-2_1, © Springer-Verlag Berlin Heidelberg 2012

opments with respect to the challenges posited by the heterogeneity and multitude of linguistic resources available today. Many of these approaches follow the **Linked Data paradigm** (Berners-Lee, 2006, see below) that postulates rules for the publication and representation of web resources. If (linguistic) resources are published in accordance with these rules, it is possible to follow links between existing resources to find other, related data and exploit network effects.

This volume provides an overview of the broad variety of approaches towards the application of the Linked Data paradigm to linguistic resources. It assembles the contributions of the workshop on Linked Data in Linguistics (LDL-2012), held at the 34th Annual Meeting of the German Linguistic Society (Deutsche Gesellschaft für Sprachwissenschaft, DGfS), March 7th-9th, 2012, in Frankfurt/M., Germany, organized by the Open Linguistics Working Group (OWLG)[1] of the Open Knowledge Foundation (OKFN),[2] an initiative of experts from different fields concerned with linguistic data, including academic linguists (e.g., typology, corpus linguistics), applied linguistics (e.g., computational linguistics, lexicography and language documentation), and NLP engineers (e.g., from the Semantic Web community). The primary goal of the working group is to promote the idea of open linguistic resources, to develop means for their representation, and to encourage the exchange of ideas across different disciplines. Accordingly, the current volume represents a great band-width of contributions from various fields, representing principles, use cases, and best practices for using the Linked Data paradigm to represent, exploit, store, and connect different types of linguistic data collections.

One goal of this book and the workshop on Linked Data in Linguistics (LDL-2012) is to document and to summarize these developments, and to serve as a point of orientation in the emerging domain of research on Linked Data in Linguistics. This documentary goal is complemented by social goals: (a) to facilitate the communication between researchers from different fields who work on linguistic data within the Linked Data paradigm; and (b) to explore possible synergies and to build bridges between the respective communities, ranging from academic research in the fields of language documentation, typology, translation studies, digital humanities in general, corpus linguistics, computational lexicography and computational linguistics, and computational lexicography to concrete applications in Information Technology, e.g., machine translation, or localization.

2 Technological Background

Several standards developed by different initiatives are referenced or used throughout this work. One is the **Extensible Markup Language** (XML, Bray et al., 1997) and its predecessor, the Standard Generalized Markup Language (SGML, Goldfarb and Rubinsky, 1990). These are text-based formats that allow to encode documents

---

[1] http://linguistics.okfn.org
[2] http://okfn.org

in an appropriate way for representing and transmitting machine-readable information.

XML and SGML have been the basis for most proposals for **interoperable representation formalisms specifically for linguistic resources**, for example the Corpus Encoding Standard (CES, Ide, 1998) developed by the Text Encoding Initiative (TEI),[3] or the Graph Annotation Format (GrAF, Ide and Suderman, 2007) developed in the context of the Linguistic Annotation Framework (LAF) by ISO TC37/SC4.[4] Earlier standards for linguistic corpora used XML data structures (i.e., trees) directly, but since Bird and Liberman (2001), it is generally accepted that generic formats to represent linguistic annotations should be based on graphs. State-of-the-art formalisms for linguistic corpora follow this assumption, and represent linguistic annotations in XML standoff formats, i.e., as bundles of XML files that are interlinked with cross-references, e.g., with formats like ATLAS (Bird and Liberman, 2001), PAULA XML (Dipper, 2005), or GrAF (Ide and Suderman, 2007).

In parallel to these formalisms, which are specific to linguistic resources, other communities have developed the **Resource Description Framework** (RDF, Lassila and Swick, 1999). Although RDF was originally invented to provide formal means to describe resources, e.g. books in a library or in an electronic archive (hence its name), its data structures were so general that its use has extended far beyond the original application scenario. RDF is based on the notion of **triples** (or 'statements'), consisting of a **predicate** that links a **subject** to an **object**. In other words, RDF formalizes relations between resources as labeled edges in a directed graph. Subjects are represented using globally unique Uniform Resource identifiers (URIs) and point (via the predicate) to another URI, the object part, to form a graph. (Alternatively, triples can have simple strings in the object part that annotate the subject resource.) At the moment, RDF represents the primary data structure of the Semantic Web, and is maintained by a comparably large and active community. Further, it provides crucial advantages for the publication of linguistic resources in particular: RDF provides a graph-based data model as required by state-of-the-art approaches on generic formats for linguistic corpora, and several RDF extensions were specifically designed with the goal to formalize knowledge bases like terminology data bases and lexical-semantic resources. For resources published under open licenses, an RDF representation yields the additional advantage that resources can be interlinked, and it is to be expected that an additional gain of information arises from the resulting network of resources. If modeled with RDF, linguistic resources are thus not only **structurally interoperable** (using RDF as representation formalism), but also **conceptually interoperable** (with metadata and annotations are modeled in RDF, different resources can be directly linked to a single repository). Further, concrete applications using linguistic resources can be build on the basis of the rich ecosystem of format extensions and technologies that has evolved around RDF, including APIs, RDF databases (triple stores), the query language SPARQL, data browsing and visualization tools, etc.

---

[3] http://www.tei-c.org
[4] http://www.tc37sc4.org

For the formalization of knowledge bases, several RDF extensions have been provided, for example the **Simple Knowledge Organization System** (SKOS, Miles and Bechhofer, 2009), which is naturally applicable to lexical-semantic resources, e.g., thesauri. A thorough logical modeling can be achieved by formalizing linguistic resources as ontologies, using the **Web Ontology Language** (OWL, McGuinness and Van Harmelen, 2004), another RDF extension. OWL comes in several dialects (profiles), the most important being OWL/DL and its sublanguages (e.g. OWL/Lite, OWL/EL, etc.) that have been designed to balance expressiveness and reasoning complexity (McGuinness and Van Harmelen, 2004; W3C OWL Working Group, 2009) OWL/DL is based on Description Logics (DL, Baader et al., 2005) and thus corresponds to a *decidable* fragment of first-order predicate logic. A number of reasoners exist that can draw inferences from an OWL/DL ontology and verify consistency constraints. Primary entities of OWL Ontologies are **concepts** that correspond to classes of objects, **individuals** that represent instances of these concepts, and **properties** that describe relations between individuals. Ontologies further support **class operators** (e.g. intersection, join, complement, instanceOf, subClassOf), as well as the specification of **axioms** that constrain the relations between individuals, properties and classes (e.g. for property P, an individual of class A may only be assigned an individual of class B). As OWL is an extension of RDF, every OWL construct can be represented as a set of RDF triples.

RDF is based on globally unique and accessible URIs and it was specifically designed to establish links between such URIs (or resources). This is captured in the **Linked Data paradigm** (Berners-Lee, 2006) that postulates four rules:

1. Referred entities should be designated by URIs,
2. these URIs should be resolvable over HTTP,
3. data should be represented by means of standards such as RDF,
4. and a resource should include links to other resources.

With these rules, it is possible to follow links between existing resources to find other, related, data and exploit network effects. The **Linked Open Data (LOD) cloud**[5] represents the resulting set of resources. If published as Linked Data, linguistic resources represented in RDF can be linked with resources already available in the Linked Open Data cloud. At the moment, the LOD cloud covers a number of lexico-semantic resources, including the Open Data Thesaurus,[6] WordNet,[7] Cornetto (Dutch WordNet),[8] DBpedia (machine-readable version of the Wikipedia),[9] Freebase (an entity database),[10] OpenCyc (database of real-world con-

---

[5] http://lod-cloud.net

[6] http://vocabulary.semantic-web.at/PoolParty/wiki/OpenData

[7] http://semanticweb.cs.vu.nl/lod/wn30,
http://www.w3.org/TR/wordnet-rdf,
http://wordnet.rkbexplorer.com

[8] http://www2.let.vu.nl/oz/cltl/cornetto

[9] http://www.dbpedia.org

[10] http://freebase.com

cepts),[11] and YAGO (a semantic knowledge base).[12] Additionally, the LOD cloud includes knowledge bases of information about languages and bibliographical information that are relevant for here, e.g., Lexvo (metadata about languages),[13] lingvoj (metadata about language in general),[14] Project Gutenberg (bibliographical data base)[15] and the OpenLibrary (bibliographical data base).[16] Given the interest that researchers take in representing linguistic resources as Linked Data, continuing growth of this set of resources seems to be assured. Several contributions assembled in this volume discuss the linking of their resources with the Linked Open Data cloud, thereby supporting the overarching vision of a Linguistic Open Data (sub-) cloud of linguistic resources, a **Linguistic Linked Open Data cloud** (LLOD).

The workshop on Linked Data in Linguistics (LDL-2012) was the first major event organized by the Open Linguistics Working Group (OWLG), and we hope that the workshop and this volume contribute to the on-going formation of an interdisciplinary community actively working towards the application of the Linked Open Data paradigm to all forms of linguistic resources, that they facilitate the exchange of ideas and resources with the long-term goal to build such a Linguistic Linked Open Data cloud.

One goal of this book is to document and to summarize recent developments in this direction, and to serve as a point of orientation to where research on Linked Data in Linguistics is heading to. Almost more important, however, is the second goal we pursued, i.e., to facilitate the communication between researchers working in this direction, to explore possible synergies and to build bridges between these communities, and we would like to thank the participants of the Linked Data in Linguistics workshop as well as the members of the OWLG for sharing their ideas, insights and/or resources, and we hope that, together, we can build a Linked Data (sub)cloud of linguistic resources that can be used across different disciplines for mutual benefit of researchers and the developers of NLP and Semantic Web applications.

2.1 Notational Conventions

Throughout this volume, the following notational conventions are applied:

- linguistic examples are written in a *slanted font*, translations are marked with single quotes
- ontological concepts, source code, URLs and URIs are written in a `typewriter font`

---

[11] http://sw.opencyc.org

[12] http://mpii.de/yago

[13] http://www.lexvo.org

[14] http://www.lingvoj.org

[15] http://www4.wiwiss.fu-berlin.de/gutendata

[16] http://openlibrary.org

• RDF literals are marked by single or double quotes

3 Structure of this Volume

We are happy to have attracted a large number of high quality contributions from very different domains for the workshop on Linked Data in Linguistics (LDL-2012) held March 7th - 9th, 2012, as part of the 34th Annual Meeting of the German Linguistics Society (DGfS) in Frankfurt a. M., Germany. The set of subdisciplines included in this volume is diverse; the goal is the same: provide scientific data in an open format which permits integration with other data repositories.

This volume is organized in four parts: Parts I, II and III describe applications of the Linked Data paradigm to major types of linguistic resources, i.e., **lexical-semantic resources**, **linguistic corpora** and **other knowledge bases**, respectively. These parts represent the contributions of the participants of the Workshop Linked Data in Linguistics (LDL-2012). In Part IV, the editors describe recent efforts to **link linguistic resources** – and thus to create a Linked Open Data (sub-)cloud of linguistic resources – in the context of the Open Linguistic Working Group (OWLG) of the Open Knowledge Foundation (OKFN). They illustrate how lexical-semantic resources, corpora and other linguistic knowledge bases can be interlinked and what possible gains of information are to be expected, using representative examples for the respective classes of linguistic resources.

As we are interested in linking different language resources, it should be noted that there is a natural overlap between these categories, and therefore, many contributions could be classified under more than one category. Bouda and Cysouw (this vol.), for example, discuss not only lexical resources, but also corpus representation, and knowledge bases for linguistic metadata; Schalley (this vol.) and Declerck et al. (this vol.) describe not only linguistic knowledge bases, but also corpus data and multi-layer annotations; and the contributions by Chiarcos (this vol.), Hellmann et al. (this vol.), and Nordhoff (this vol.) that are presented in the context of linking linguistic resources, could also have been presented in the respective parts on linguistic corpora, lexical-semantic resources and other (linguistic) knowledgebases.

3.1 Lexical Resources

Part I describes the modeling of various lexical-semantic resources as illustrated for lexical-semantic resources.

Peter Bouda and Michael Cysouw describe the digitization of dictionaries, and how the elements (head words, translations, annotations) found in there can be served in a Linked Data way while at the same time maintaining access to the document in its original form. To this end, they use standoff markup, which furthermore allows the third-party annotation of their data. They also explore how these third-party annotations could be shared in novel ways beyond the normal scope of normal academic distribution channels, e.g. Twitter.

John McCrae, Elena Montiel-Ponsoda and Philipp Cimiano describe the *lemon* format that has been developed for the sharing of lexica and machine readable dictionaries. They consider two resources that seem ideal candidates for the Linked Data cloud, namely WordNet 3.0 and Wiktionary, a large document based dictionary. The authors discuss the challenges of converting both resources to *lemon*, and in particular for Wiktionary, the challenge of processing the mark-up, and handling inconsistencies and underspecification in the source material. Finally, they turn to the task of creating links between the two resources and present a novel algorithm for linking lexica as lexical Linked Data.

Axel Herold, Lothar Lemnitzer, and Alexander Geyken report on the lexical resources of the long-term project 'Digitales Wörterbuch der deutschen Sprache' (DWDS) which aims at the integration of several lexical and textual resources in order to document the German language and its use at several stages. They describe the explicit linking of four lexical resources on the level of individual articles which is achieved via a common meta-index. The authors present strategies for the actual dictionary alignment as well as a discussion of models that can adequately describe complex relations between entries of different dictionaries.

David Lewis et al. describe perspectives of Linked Data in the fields of software localisation and translation. They present a platform architecture for sharing, searching and interlinking of Linked Localisation and Language Data on the web. This architecture rests upon a semantic schema for the respective resources that is compatible with existing localisation data exchange standards and can be used to support the round-trip sharing of language resources. The paper describes the development of the schema and data management processes, web-based tools and data sharing infrastructure that use it. An initial proof of concept prototype is presented which implements a web application that segments and machine translates content for crowd-sourced post-editing and rating.

3.2 Linguistic Corpora

Part II deals with problems to create, to maintain and to evaluate linguistic corpora and other collections of linguistically annotated data. Previous research indicates that formalisms such as RDF and OWL are suitable to represent linguistic annotations (Burchardt et al., 2008; Cassidy, 2010) and to build NLP architectures on this basis (Wilcock, 2007; Hellmann, 2010), yet so far, it has rarely been applied to this type of linguistic resource.

Marieke van Erp describes interoperability problems of linguistic resources, in particular corpora, and develops a vision to apply the Linked Data approach to these issues. In her contribution, the constraints for linguistic resource reuse and the tasks are detailed, accompanied by a Linked Data approach to standardise and reconcile concepts and representations used in linguistic annotations.

As mentioned above, these problems are addressed in the NLP community by generic data models for linguistic corpora that are based on directed graphs. Kerstin Eckart, Arndt Riester and Katrin Schweitzer describe such a state-of-the-art approach on the task of resource integration for multiple independent layers of annotation in a multi-layer annotated corpus that is based on a graph-based data model, although not on RDF, but an XML standoff format and a relational database management system. They present an annotated corpus of German radio news including syntactic information from a parser, as well as manually annotated information status labels and prosodic labels. They describe each annotation layer and focus on the linking of the data from both layers of annotation, and show how the resource can support data extraction on both annotation layers. Although they do not directly make use of the Linked Data paradigm, the problems identified and the data model employed represent important steps towards the development of representation formalisms for multi-layer corpora by means of RDF and as Linked Data, see, for example, Chiarcos (this vol.).

Michael Carl and Henrik Høeg Müller describe a fascinating intersection between pure structural syntactic data and human-machine interaction in translation processes. Human behaviour while translating on a computer can be recorded with eye trackers and capturing of user input (mouse, keyboard). This behavioural data can then be linked to syntactic data extracted from the sentence translated (constituency, dependency). The intuition is that syntactically complicated sentences will have a repercussion in the user behaviour (longer gaze, slower input, more corrections). Carl and Müller, just like Bouda and Cysouw, and Eckart et al., use standoff annotation to allow for overlapping annotations. Their use of structural data on the one hand and behavioural data from a novel domain on the other hand shows the benefits the provision of data as Linked Data can have.

María Blume, Suzanne Flynn and Barbara Lust describe DTA, an online tool for the study of language acquisition. DTA allows for data creation, data management and collaborative use of child language data from a variety of languages (Spanish, French, English, Sinhala). Language Acquisition is a relative newcomer to the area of Linked Data, and it is exciting to see that areas somewhat distant from the NLP origins of Linked Data are beginning to join the movement.

3.3 Linguistic Knowledgebases

While Part II focused on annotated linguistic data, Part III presents a number of repositories of knowledge about languages and linguistic terminology that can be used, for example, for annotating linguistic data with linguistic analyses and metadata.

Menzo Windhouwer and Sue Ellen Wright describe the linking from language resources to linguistic data categories in ISOcat, a repository of linguistic terminology developed to foster semantic interoperability of linguistic resources. This registry follows a grass roots approach, which means that any linguist can add the data categories (s)he needs. However, the goal of improving semantic interoperability can only be met if the data categories are reused by a wide variety of linguistic resource types. A resource indicates its usage of data categories by linking to them, this paper describes the technical prerequisites to achieve this in an RDF-based approach.

Thierry Declerck et al. describe strategies for exploiting the large set of dynamically increasing, freely available language data incorporated in the Linked Open Data (LOD) framework. Such language data currently mostly exist in the form of raw, unstructured textual expressions within RDF labels or comments. Incorporating them as structured language data within the LOD leads to a linguistic enrichment of the data sets that express linked (domain) knowledge resources, and this will enable the creation of more accurate, knowledge-aware NLP applications. This integration of linguistic information in knowledge representation systems should be done in compliance with both ISO (multi-layer linguistic annotation and data categories) and W3C (RDF, SKOS) standards. By this, new linguistically enriched datasets can also be more easily ported into the LOD format: e.g., repositories in the field of Digital Humanities often hold language data in taxonomical structures. The potential of linked language data for digital humanities is illustrated here for the detection of motifs in literary texts. For this purpose, a formal representation of the taxonomical structure of the Thompson Motif-Index of folk-literature (Thompson, 1955-58) is presented.

In a similar vein, Antonia Pareja-Lora reports on the development of a concept taxonomy for a different type of linguistic annotation, namely pragmatic annotations. Pragmatics has to deal with a real mix of different linguistic topics, such as (i) speech acts, (ii) deixis, presuppositions and implicatures; or (iii) pragmatic coherence relations, which traditionally have been tackled following several fragmentary and/or partial approaches. Pareja-Lora describes an approach to specify formally the different elements that a pragmatic annotation scheme should contemplate and make explicit with the goal to facilitate the interoperability of linguistic annotations up to the pragmatic level.

While the terminology repositories and taxonomies described in this part so far have been developed for interoperability of NLP tools and linguistic annotations, the remaining chapters of this part deal with typological databases that provide information about languages from a slightly different angle of research.

Steve Moran tackles the very basic unit of linguistics, the phoneme, and shows how heterogeneous data bases of phoneme inventories found in the worlds languages can be integrated with a Linked Data approach via mapping of the relations found in the original data bases to his ontology. His system is in production stage, and Moran shows how a number of phonological hypotheses can be confirmed or refuted using his PHOIBLE database. Moran furthermore explores the difference between queries in traditional relational databases and SPARQL queries.

Andrea Schalley casts a wide net and lists the criteria a typological knowledge base would have to respond to in an ideal world. She then discusses challenges for the realization and sketches the development of a computational tool that utilises Semantic Web technologies in order to provide novel ways to process, integrate, and query cross-linguistic data. Its data store incorporates a set of ontologies (comprising linguistic examples, annotations, language background information, and metadata) backed by a software logic reasoner. This allows for highly targeted querying and answers on rather specific questions such as (i) which size (in terms of speaker count) do languages have that have kin-sensitive pronouns?, or (ii) which languages code joint attention in their grammar, and if so, where in the grammar do they do it?

3.4 Towards a Linguistic Linked Open Data Cloud

The last part describes joint activities of different members of the Open Linguistics Working Group (OWLG) aiming to develop a Linked Open Data (sub-)cloud of linguistic resources.

Christian Chiarcos, Sebastian Hellmann and Sebastian Nordhoff describe the Open Linguistics Working Group (OWLG), its goals, addressed problems, recent activities and on-going developments.

Christian Chiarcos describes the formalization of annotated linguistic corpora by means of OWL/DL with a focus on genericity and interoperability. Structural interoperability of linguistic corpora is addressed with POWLA, an OWL/DL formalization of a data model designed to represent any kind of linguistic annotation assigned to textual data; conceptual interoperability between annotations of different corpora can be established using the OLiA ontologies, an architecture of modular OWL/DL ontologies that formalize the linking of annotation schemes with community-maintained terminology repositories.

Sebastian Hellmann, Claus Stadler and Jens Lehmann describe the DBpedia, one of the major free data sets in the Web of Data, as an example of a lexical-semantic resource. In particular, the internationalization of the DBpedia is addressed – including the development of a German DBpedia. The authors further describes the NLP Interchange Format (NIF), that can be used, for example, to develop NLP pipelines that perform the task to assign words the corresponding DBpedia concept (entity linking). NIF represents the output of NLP tools in RDF, and thus, it is possible to integrate this data into an existing Linked Data infrastructure.

Sebastian Nordhoff presents a knowledge base that conveys information *about* linguistic resources, it thus exemplifies how metadata can be provided within the Linguistic Linked Open Data cloud: Sebastian Nordhoff describes how existing work on language classification can interface with bibliographical work based on standards like TEI and Dublin Core in the Glottolog/Langdoc project. His work affords links to the vast amounts of bibliographical data contained in the LOD cloud on the one hand, and language classification and language history on the other. Further, he illustrates the linking between LOD resources for the example of Glottlog/Langdoc and ASJP online, which measures the lexical distance between languages.

Using POWLA, the DBpedia, OLiA and Glottolog/Langdoc as examples, the final contribution by Christian Chiarcos, Sebastian Hellmann and Sebastian Nordhoff describes how corpora, lexical-semantic resources, and other linguistic knowledge bases can be interlinked, and how additional information can be obtained by building a Linked Open Data (sub-)cloud of linguistic resources.

References

Baader F, Horrocks I, Sattler U (2005) Description logics as ontology languages for the Semantic Web. Mechanizing Mathematical Reasoning pp 228–248

Berners-Lee T (2006) Design issues: Linked data. `http://www.w3.org/DesignIssues/LinkedData.html`

Bird S, Liberman M (2001) A formal framework for linguistic annotation. Speech Communication 33(1-2):23–60

Bouda P, Cysouw M (this vol.) Treating dictionaries as a Linked-Data corpus. pp 15–23

Bray T, Paoli J, Sperberg-McQueen C, Maler E, Yergeau F (1997) Extensible Markup Language (XML). World Wide Web Journal 2(4):27–66

Burchardt A, Padó S, Spohr D, Frank A, Heid U (2008) Formalising Multi-layer Corpora in OWL/DL – Lexicon Modelling, Querying and Consistency Control. In: Proceedings of the 3rd International Joint Conference on NLP (IJCNLP 2008), Hyderabad

Cassidy S (2010) An RDF realisation of LAF in the DADA annotation server. In: Proceedings of the 5th Joint ISO-ACL/SIGSEM Workshop on Interoperable Semantic Annotation (ISA-5), Hong Kong

Chiarcos C (this vol.) Interoperability of corpora and annotations. pp 161–179

Declerck T, Lendvai P, Mörth K, Budin G, Váradi T (this vol.) Towards Linked Language Data for Digital Humanities. pp 109–116

Dipper S (2005) XML-based stand-off representation and exploitation of multi-level linguistic annotation. In: Proc. Berliner XML Tage 2005 (BXML 2005), Berlin, Germany, pp 39–50

Dostert L (1955) The Georgetown-IBM experiment. In: Locke WN, Booth AD (eds) Machine Translation of Languages, John Wiley & Sons, New York, pp 124–135

Francis WN, Kucera H (1964) Brown Corpus manual. Manual of information to accompany A standard corpus of present-day edited American English, for use with digital computers. Tech. rep., Brown University, Providence, Rhode Island, revised edition 1979

Goldfarb CF, Rubinsky Y (eds) (1990) The SGML handbook. Oxford University Press, New York

Hellmann S (2010) The semantic gap of formalized meaning. In: The 7th Extended Semantic Web Conference (ESWC 2010), Heraklion, Greece

Hellmann S, Stadler C, Lehmann J (this vol.) The German DBpedia: A sense repository for linking entities. pp 181–189

Ide N (1998) Corpus Encoding Standard: SGML guidelines for encoding linguistic corpora. In: Proceedings of the First International Language Resources and Evaluation Conference (LREC 1998), pp 463–70

Ide N, Pustejovsky J (2010) What does interoperability mean, anyway? Toward an operational definition of interoperability. In: Proc. Second International Conference on Global Interoperability for Language Resources (ICGL 2010), Hong Kong, China

Ide N, Suderman K (2007) GrAF: A graph-based format for linguistic annotations. In: Proc. Linguistic Annotation Workshop (LAW 2007), Prague, Czech Republic, pp 1–8

Lassila O, Swick RR (1999) Resource Description Framework (RDF) model and syntax specification. Tech. rep., World Wide Web Consortium

McGuinness D, Van Harmelen F (2004) OWL Web Ontology Language overview. w3c recommendation. Tech. rep., World Wide Web Consortium

Miles A, Bechhofer S (2009) SKOS Simple Knowledge Organization System reference. W3C Recommendation. Tech. rep., World Wide Web Consortium

Morris W (ed) (1969) The American Heritage Dictionary of the English Language. Houghton Mifflin, New York

Nordhoff S (this vol.) Linked Data for linguistic diversity research: Glottolog/Langdoc and ASJP. pp 191–200

Schalley AC (this vol.) TYTO – A collaborative research tool for linked linguistic data. pp 139–149

Text Encoding Initiative (1990) TEI P1 guidelines for the encoding and interchange of machine readable texts. Tech. rep., Text Encoding Initiative, draft Version 1.1 1

Thompson S (1955-58) Motif-index of folk-literature: A classification of narrative elements in folktales, ballads, myths, fables, medieval romances, exempla, fabliaux, jest-books, and local legends. Indiana University Press, Bloomington

W3C OWL Working Group (2009) OWL 2 Web Ontology Language. document overview. W3C Recommendation. Tech. rep., World Wide Web Consortium

Wilcock G (2007) An OWL ontology for HPSG. In: Proc. 45th Annual Meeting of the Association for Computational Linguistics, Prague, Czech Republic, pp 169–172

Part I
Lexical Resources

Treating Dictionaries as a Linked-Data Corpus

Peter Bouda and Michael Cysouw

Abstract In this paper we describe a practical approach to the challenge of linguistic retrodigitization. We propose to distinguish strictly between a base digitization and separate interpretation of the sources. The base digitization only includes a literal electronic transcript of the source. All sources are thus simply treated as strings of characters, i.e. as unstructured corpora. The often complex structure as found in many dictionaries and grammars will subsequently (and possibly much later) be added as Linked Data in the form of standoff annotation. A further advantage of this approach is that the complete digitization and interpretation can be performed collaboratively without a complex organizational superstructure.

1 Introduction

A large amount of the knowledge about the world's languages is currently only available in traditionally printed form, as grammars, text collections or dictionaries. Although this body of knowledge is large, it is finite and manageable in size given current computational power and electronic storage. A proper retrodigitization of these resources would allow for many new approaches to the quantitative comparison of languages, be it for a better understanding of cross-linguistic variation in grammatical structure or for new and improved historical-comparative reconstructions.

Still, the number of pages to be digitized is large enough to pose serious challenges for the organizational infrastructure (we estimate the number of pages to be digitized for the world's lesser-studied languages to be in the order of 10^6). Not only the size, but also the necessary precision of the digitization poses special desiderata. To allow for proper linguistic analysis, the precision of the digitization has to

Peter Bouda · Michael Cysouw
Research Unit "Quantitative Language Comparison", Ludwig Maximilians University Munich
e-mails: pbouda@cidles.eu, cysouw@lmu.de

C. Chiarcos et al. (eds.), *Linked Data in Linguistics*,
DOI 10.1007/978-3-642-28249-2_2, © Springer-Verlag Berlin Heidelberg 2012

be highly accurate, because linguistic description is a strongly technical tradition in which each dot, dash, and tilde, and all italics, boldfaces and tab-marks have a specific and important meaning – and unfortunately a different meaning in each source.

This digitization will probably never be perfect the first time round. Also, the interpretation of all the special symbols used will be a task to be handled in many years to come, long after the basic digital encoding of the sources has been completed. The real challenge of linguistic retrodigitization is thus not the digital encoding as such, but the continuing update, refinement, and interpretation of the digital products.

In our view it is of central importance that everybody working with the digitized data should always able to trace back the information to the original source. Further, it should be possible to reconstruct every step in the digitization workflow to make it possible to find and correct errors. We propose a framework that allows scientists to work and enrich the digital data while maintaining this traceability. To accomplish this, we describe several technical and architectural solutions we devised in our project in which we are retrodigitizing dictionaries. What we create is a new type of linked-data corpus that is derived from legacy printed material and that generates new opportunities in research for a global scientific community, if done right.

2 Base Digitization and Annotation

The first step in the digitization process is the scanning and transcription of printed dictionaries. The end product of this transcription process is a basic text document with typographic and layout information. This transcript is transformed into an XML document which is the basis for the subsequent processing steps, which we describe in this paper. This raw digital version should minimally have basic formatting tags mimicking the printed original (i.e. italic, bold, etc), the original line breaks and indentation, and information about page and column numbers. From this information it is possible to approximately (though not necessarily perfectly) recreate the text as it looks in the original printed source.

Most importantly, this raw digital version does not include any interpretation about what the structure of the printed original is supposed to mean. For example, many dictionaries use italics to signify structure (e.g. parts of speech, or examples), but this structure will only be added later as Linked Data, so differing interpretations are possible. For such a interlinked structure to remain intact, the base version has to be as static and persistent as possible. So, we prefer a maximally simple base digitization as a start.

2.1 Basic Chunking

To ease annotation, the whole document is separated into reasonable and manageable chunks. Those chunks should be small enough to allow annotation through character counts. Although character counts could of course just as well work with complete books, for reasons of error correction and traceability we prefer chunks not larger than about a thousand characters (i.e. paragraph size). The chunks should also not be too small, so as to allow a human reader to quickly understand what she is looking at. Again, this is purely for reasons of manageability. In retrodigitization, we think that it is important not only to consider technical considerations, but also include arguments pertaining to social management and human interfaces. So we propose not to use word chunking, or even more complex linguistically-based chunking on the basic digitalization level. Further, also for reasons of traceability, those chunks should preferably be derived from the inherent structure of the sources. In our case of printed dictionaries we decided to use the entries of our dictionaries as basic tokens. For other sources, any available paragraph structure can be used to define chunks.

It is also necessary to remember the page number for each chunk, as we want to be able to approximately reconstruct the original printed pages. As the unique ID of each chunk, we use a human readable description which consists of two parts: the (start) page number and the relative position on the page. Table 1 shows one example entry and the information we store for it in the base digitization.

Table 1 A dictionary entry in the base digitization.

Field	Value
fullentry	afebeba (s1B) cuarto de arriba.
start page	22
start column	1
position on page	2

2.2 Adding Source Information as Annotations

We prefer a base digitization that does not include any internal structure except for the linear structure of the text, as this makes their handling much easier later on. So, all formatting information that is present in the original XML transcript is removed. This information is stored as Linked Data in the form of standoff annotations. The annotation refers to entries via its ID and by using character counts. For example, one annotation could have the information that the entry /5/10/ (i.e. the one on page 5 at position 10) has italic characters from character position 9 to 14, another annotation contains a newline at position 23. Table 2 shows one such annotation as an example.

Table 2 Data fields for annotation.

Field	Value
type	pagelayout
value	newline
start	23
end	23

Our annotation tools also alter the original transcript in one important way: we remove hyphens before line breaks and instead store an annotation called hyphen for the position where we removed it. Hyphenation is not considered to be content-related information, but only induced by the printed format of the original. We remove it from the base digitization because it is not necessary for later interpretation. For reasons of traceability, we keep the information, so we can reconstruct the original using this annotation.

2.3 Step-by-Step Enrichment

One of the advantages of this basic division into 'flat' base digitization and standoff annotations is the possibility to add information step-by-step by adding further annotations. For example, in our project the prime emphasis is on using the head words and translations of the dictionaries. We are able to find this information easily within the entries. Often, head words are printed in special format (bold or italic) and translations start and end after or before certain characters. We then save this induced information in the same way as the annotations mentioned in the previous paragraph, i.e. by adding exactly the same kind of standoff annotations.

Many of our dictionaries contain additional information about part-of-speech, some of them also have phonological and morphological descriptions of head words. There are often example sentences with translations, and all kind of further information. We do not parse all this information right now, as we do not need it for the current project. But other researchers can extract this information if wanted. In simple cases we already add additional annotations, but in other cases we leave this task to future research projects that might be interested in different information provided in the sources. The basic structure of our corpus allows us to focus on the things we need right now, but still open up the possibility of enriching the data in the future. No matter whether we do it ourselves or someone else adds interpretations.

3 (Re-)Publication and Collaboration

In addition to the internal work with the sources within our project, our goal is to publish the digitized dictionaries as a corpus. The data should be free to use by anyone (pending copyright issues). Again, there are two main principles our corpus structures needs to fulfill. On one hand, every researcher should be able to trace back everything we did with the data, up to the original entry, on the original page in the printed dictionary. Further, when a researcher thinks our annotations are not good enough, or wants to add information to our annotations, she should simply be able to do this, in an easy and independent way. Our framework proposes several means how to follow these principles, namely the usage of standoff annotations, an XML format and a certain URL structure to maintain traceability even all the way to using a printed URL.

3.1 Standoff Annotations

Storing information as standoff annotations has several advantages. First, users can just download the type of data they need. If someone needs plain dictionary entries, then she downloads the basic data file. If she needs additional information about line breaks and indentation, she downloads another file. This modularization makes data handling easier, even more so when more and more layers of annotations are added.

Second, the basic data has only the information one needs for automated linguistic analysis. It contains plain stings stripped of any structural information. Standards like those of the Text Encoding Initiative[1] (TEI) propose to store formatting, layout and structural information (especially for dictionaries) within the basic data. In our view, this leads to problems when later enriching the data with additional annotations. It is not very clear how tokenization and standoff annotation should work when tags are used extensively inside the basic data (cf. Cayless and Soroka, 2010; Bánski and Przepiórkowski, 2009).

Third, using a simple yet powerful and far-reaching standoff annotation from the beginning allows us to collaborate with other scientists, for example specialists for the language families we work with. Researchers can just add annotations without the need to remove or alter any of our annotations. It is also easily possible to integrate changes of the data with our annotations, if we want to permanently store them. But this integration is always optional: if we or someone else wants to publish personal interpretations of certain annotations she is free to do so. Here, the basic entry with a fixed ID serves as a reference point that connects all linked annotations.

---

[1] http://www.tei-c.org/release/doc/tei-p5-doc/en/html/index.html

3.2 XML Format

XML data can sometimes be hard to handle, especially if you have large files and complex structure. There are cases where researchers prefer to have plain text files to process data (Schmidt, 2010). In our case, we prefer an XML structure that only has a minimum complexity, but still can represent every information we have. Given that our basic chunks are just strings, and annotations have only few data fields to store, the resulting XML is easily manageable. We are currently using a XML format that is derived from the proposals of the Corpus Encoding Standard[2] (CES), an early application of the TEI standard (see listing 1). We are aware that this standard is not actively developed anymore, and that the TEI is working on new standards that should also fulfill our needs in the end (Lee and Romary, 2010). But right now we see the CES-XML as the best way to store and exchange data like ours. CES is very easy to read, the specification is quite clear and focussed on the usability of data in different environments. Given the simple data structure of our work, any transformation into different XML structures should be trivial.

In general, though, the structure in which the data is stored is just a collection of linked-data entities, so everything is perfectly compatible with a more forward-looking RDF approach. For reasons of practical manageability we have decided not to use RDF as the underlying data model, but (for the moment) to rely on more traditional concepts like data tables with types and values. Conceptually, it is trivial to transform our data into an RDF representation, but the practical effort involved has kept us from providing such an access to our data right now. That will be done in the near future.

Listing 1 Example XML snippet for dictionary entry.

```
<div type="dictentry">
  <p id="22.2">
    afebeba (s1B) cuarto de arriba.
  </p>
</div>
<chunk from="22.2/0">
  <tok type="pagelayout" value="newline"
       from="22.2/23" to="22.2/23">
    <orth></orth>
  </tok>
</chunk>
```

3.3 URLs as Source Pointers

URLs are one of the most important means to publish and exchange scientific research nowadays, yet still most URLs give no hint on what kind of data is available behind them. In digital archives the URLs unfortunately often only con-

---

[2] http://www.cs.vassar.edu/CES/

tain numerical IDs. We want to use our URLs as references to a web page, but also to the original source. It should be usable for online needs, but also in printed publications. A reader who reads an article about our data should be able to take the dictionary from her shelf and look up the entry we discuss in the paper. As a result, our URLs contain a transparent string ID for the book, page numbers and the position on the page for entries and their annotations. A link to the entry page of a dictionary part of a book, for example, looks like this: `/source/thiesen1998/dictionary-25-339.html`. In this case, `thiesen1998` is our ID for Thiesen and Thiesen (1998), the dictionary part begins on page 25 and ends on page 339 of the book. The following two URLs reference smaller parts of it, for example all entries on page 25: `/source/thiesen1998/25/index.html`, or the 9th entry on page 25 and its annotations: `/source/thiesen1998/25/9/index.html`.

This structure will also be preserved when we transfer the data to any file-based archive. In a file-based archive, there will be a "main" folder for each dictionary (called `thiesen1998` in this case) and several sub-folders for pages and entries. This mapping of URLs to a files-and-folders structures and vice versa reduces the costs of data handling (as one can simply mirror our website to have a full archive structure) and allows easy traceability of every reference that will be published in the future, back to a web page or an archive folder, and even to the original dictionary.

3.4 Tags as Source Pointers

In addition to the URLs we propose another, more general possibility to refer to original sources and to cite in online publications. This technique is derived from tagging facilities in blogging and micro-blogging systems. The tags there are normally used to group the entries of one or more blogs under certain keywords, for example the tag `Linguistics` or hashtag `#linguistics` is used to group all blog articles, tweets, etc. that have linguistic content. We propose to adapt this procedure to allow scientist refer to the sources in an easy and intuitive way. The basic idea is to have a special tag (we propose `litref` to differentiate from existing tags) and add specifications about book, pages, URLS, etc. separated by slashes. If a scientist wants to refer to page 25 of the book (Thiesen and Thiesen, 1998) then a hashtag might look like this: `#litref/thiesen1998/page25`. To be more specific a ISBN or OCLC code may be used: `#litref/oclc/40505215`. The format of the tag should be as free as possible, as the most important thing is that scientists can cite in their electronic publications as easy as (or even easier than) in printed articles.

The next step is then to search and index those tags from all web pages, blogging and micro-blogging hosts. This task can be done by an adapted search robot that parses the structure of the tag and tries to find the source in a bibliographical database. Tags with ISBN, ISSN or some other codes are easy to parse, tags with references like `thiesen1998` might require some heuristics but should pose no bigger problem to state-of-the-art search robots. In the end the database consists of the bibliographical entry plus all the web pages that refer to that entry and possibly additional information like page and line numbers. This results in a huge network of Linked Data that requires nothing more than users who agree on a certain tag and its template. Existing infrastructure like blogs and search robots can be used to create such a network.

Bibliographical entries are of course not restricted to printed publications. If researchers want to cite an electronic article (like a blog entry) they might just use the same tagging mechanism with a tag like `#litref/url/http://www...`; or even without the term `url` because it may be derived by the information given by the prefix `http://`. Another possible addition to the proposal is the introduction of a new (X)HTML meta-tag that contains all the tags for the sources that the current web page refers to. This is easy to do when a web page contains only one or two tags and makes it easier for search robot to harvest the tags and integrates this approach into the broader context of the semantic web. The whole infrastructure should still work without these meta-tags, it is an optional addition to the proposal.

A use case for such a framework in our project is a crowdsourcing approach for the corrections of dictionary entries and annotations. Possible co-workers may use a service like twitter to publish corrections to certain entries by using the proposed tags. A search robot than collects those tags and adds the content of the tweets to the entry in our database and on our website. We later manually apply the proposed changes to integrate the corrections into our database.

4 Summary

The combination of stable source URLs and the standoff annotation pointers provide a stable and easily manageable infrastructure for retrodigitization. As long as the source is kept simple and stable, multiple independent annotations can be added without the need for a central infrastructure, thus allowing collaborative annotation of the important and rich source of our linguistic heritage.

References

Bánski P, Przepiórkowski A (2009) Stand-off TEI annotation: The case of the National Corpus of Polish. In: Proceedings of the Third Linguistic Annotation Workshop (LAW III), pp 65–67

Cayless HA, Soroka A (2010) On implementing string-range() for TEI. In: Proceedings of Balisage: The Markup Conference 2010

Lee K, Romary L (2010) Towards interoperability of ISO standards for Language Resource Management. In: Proceedings of the 2nd International Conference on Global Interoperability for Language Resources

Schmidt D (2010) The inadequacy of embedded markup for cultural heritage texts. Literary and Linguistic Computing pp 337–356

Thiesen W, Thiesen E (1998) Diccionario Bora-Castellano Castellano-Bora. Instituto Lingüístico de Verano

Integrating WordNet and Wiktionary with *lemon*

John McCrae, Elena Montiel-Ponsoda, and Philipp Cimiano

Abstract Nowadays, there is a significant quantity of linguistic data available on the Web. However, linguistic resources are often published using proprietary formats and, as such, it can be difficult to interface with one another and they end up confined in "data silos". The creation of web standards for the publishing of data on the Web and projects to create Linked Data have lead to interest in the creation of resources that can be published using Web principles. One of the most important aspects of "Lexical Linked Data" is the sharing of lexica and machine readable dictionaries. It is for this reason, that the *lemon* format has been proposed, which we briefly describe. We then consider two resources that seem ideal candidates for the Linked Data cloud, namely WordNet 3.0 and Wiktionary, a large document based dictionary. We discuss the challenges of converting both resources to lemon , and in particular for Wiktionary, the challenge of processing the mark-up, and handling inconsistencies and underspecification in the source material. Finally, we turn to the task of creating links between the two resources and present a novel algorithm for linking lexica as lexical Linked Data.

1 Introduction

In the last decade, a large amount of linguistic and lexical resources in particular have been created. These resources are confined however to what Tim Berners-Lee has named "data silos", as either they are publicly available, albeit in proprietary formats, or access to them is restricted. This leads to a situation in which the in-

John McCrae · Philipp Cimiano
CITEC, Universität Bielefeld, e-mail: {jmccrae,cimiano}@cit-ec.uni-bielefeld.de

Elena Montiel-Ponsoda
Ontology Engineering Group, Universidad Politécnica de Madrid, e-mail: emontiel@delicias.dia.fi.upm.es

C. Chiarcos et al. (eds.), *Linked Data in Linguistics*,
DOI 10.1007/978-3-642-28249-2_3, © Springer-Verlag Berlin Heidelberg 2012

tegration of various linguistic data becomes cumbersome. The *Linking Open Data* project (Berners-Lee, 2009) has aimed to solve these issues by fostering the publication of data on the Web using the RDF data model and, most importantly, linking data across sites. In this paper, we discuss how the principles of Linked Data can be applied to the publication of linguistic data. We discuss in detail the conversion of WordNet and Wiktionary to Linked Data resources using the *lemon* model as a use case. While WordNet has been already converted to the RDF data model, there are significant challenges in converting a semi-structured resource such as Wiktionary into the RDF data model. We discuss these challenges and how we addressed them. Our use cases demonstrate that *lemon* can be used as a uniform, principled and simple model for the publication of lexical resources as Linked Data as well as their linking. All resources described in this paper are available at `http://monnetproject.deri.ie/lemonsource`.

2 Related Work

There is a high interest in Natural Language Processing to exploit not only curated resources such as WordNet, but also collaboratively created resources such as Wiktionary[1] or Wikipedia.[2] These collaboratively created resources are especially interesting due to their coverage and due to the fact that they contain linguistic knowledge for a plethora of languages. Further, in spite of not having been created by linguists, they are still highly interesting for them as they contain huge amounts of semantically structured knowledge that is not typically available in standard linguistic resources (Zesch et al., 2008). A good example of a project integrating and linking various lexical resources is the NULEX project.[3] NULEX is a lexical resource derived automatically from from WordNet (Fellbaum, 1998), VerbNet (Kipper-Schuler, 2005) and Wiktionary. It reuses lexical information from WordNet and syntactic knowledge as well as subcategorization frames from VerbNet (an extension of the verb classification of Levin, 1993). By mapping these two resources, WordNet verbs are complemented with information about subcategorization. Finally, tense information is obtained from Wiktionary. However, the publication of linguistic resources as Linked Data does not solve the interoperability problem *per se*, as categories still need to be aligned to each other. Chiarcos has presented the OLiA framework (Chiarcos, 2010, also see Chiarcos, this vol.) for this purpose. It consists of a number of OWL DL ontologies that formalize the mapping between annotations of existing terminology repositories, such as GOLD (Farrar and Langendoen, 2003) or the ISOcat category registry (Kemps-Snijders et al., 2008, also

---

[1] `http://www.wiktionary.org`
[2] `http://www.wikipedia.org`
[3] `http://www.qrg.northwestern.edu/resources/nulex.html`

see Windhouwer and Wright, this volume). OLiA thus facilitates the mapping of various annotation schemes.

3 The *lemon* Model

lemon (LExicon Model for ONtologies, McCrae et al., in press) is an RDF model that allows to specify lexica for ontologies and allows to publish these lexica on the Web.[4] In contrast to the existing WordNet 2.0 RDF model, *lemon* is not intended to be a model for a single lexical resource, but a method by which multiple models with complementary purposes can be published, linked and shared on the Web.

The main features of the model can be summarised as follows:

- **Semantics By Reference**: Linguistic descriptions are separate from the ontology, but their semantics are defined by pointing to the corresponding semantic objects in the ontology
- **Modular Architecture:** The model consists of a core model and a set of complementary modules. Linguistic descriptions are grouped into 5 modules:

  1. Linguistic properties (e.g., part-of-speech, gender, number),
  2. Lexical and terminological variation,
  3. Decompositions of phrase structures,
  4. Syntactic frames and their mappings to the logical predicates in the ontology, and
  5. Morphological decomposition of lexical forms.

- **Openness:** *lemon* is a descriptive model that does not prescribe the usage of specific linguistic categories. Thus, the data categories or linguistic annotations used to define lexical information in the model are not captured in the *lemon* model proper, but have to be specified by reusing URIs from other dictionaries and repositories such as ISOcat or the GOLD ontology.

The core classes of the *lemon* model can be seen in Fig. 1. The core classes are the ones that form the main path between the `Ontology` and the lexical realisation represented in the `LexicalEntry` class. A `LexicalEntry` may also have multiple `LexicalForms` representing morphological variants, each of which is associated with a written representation (`writtenRep`). The `LexicalSense` class provides a principled link between an ontology concept and its lexical realization. Since 'concepts' or world objects, as defined in ontologies, and 'lexical entries', as defined in lexicons, can rarely be said to truly overlap, the `LexicalSense` class provides the adequate restrictions (usage, context, register, etc.) that make a certain lexical entry appropriate for naming a certain concept in the specific context of the ontology being lexicalised.

---

[4] Technical details of the model have been described in `http://lexinfo.net/ lemon-cookbook.pdf`.

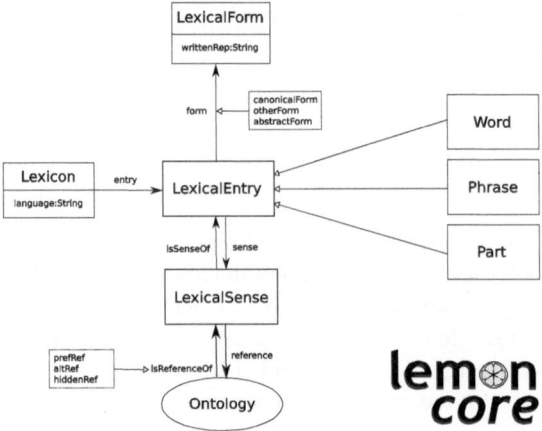

Fig. 1 The core *lemon* model

The design principles of this model make it ideal for interchanging lexica on the Web. Since *lemon* builds on the RDF data model, URIs are used to name and dereference linguistic annotations, and links can be easily created between lexicons using RDF triples. Moreover, the model is modular in the sense that, according to the final application needs, certain modules can be used or not. This also allows for new modules to be created if this is required by a certain application. In this sense, the *lemon* model can be said to be suited for the publication and linking of lexical resources on the Web.

Multilingualism is also foreseen in *lemon*, as several lexica in different languages can be associated to one and the same ontology. Moreover, translation relations can be established at the `LexicalSense` class, even allowing for conceptualization mismatches between languages to be represented, if needed. In fact, a specific module for representing translations has been proposed for *lemon* (Montiel-Ponsoda et al., 2011). The main idea of this module is to provide metadata about translations (such as provenance, confidence level, etc.), as well as to capture different types of translations (descriptive translations vs. culturally equivalent translations).

4 Methods

This section describes the methods employed to transform WordNet and Wiktionary into *lemon*, and the linking of lexical entries that are common to both resources.

4.1 WordNet

The transformation of WordNet into *lemon* has been described before by McCrae et al. (2011). This conversion was performed automatically based on the manual alignment of the WordNet vocabulary to the *lemon* vocabulary. Hereby, synsets in WordNet were essentially converted into ontology concepts, words into *lemon* lexical entries, and senses into *lemon* lexical senses, respectively. The major change was the modelling of *forms* as RDF resources, in contrast to treating them as properties. A disadvantage of using ad-hoc formats when publishing lexical resources as Linked Data is the fact that schema changes might be required when the schema of the underlying resources changes. For example, when using an ad-hoc conversion to RDF schema, the conversion of WordNet 3.0 and WordNet 2.0 would yield different schemas as form variants are specified in WordNet 3.0 in extra files. Having a principled and uniform format such as *lemon* would overcome this issue of changing RDF schemas.

4.2 Wiktionary

Wiktionary is a human-readable lexicon that is publicly available on the Web, hosted by the WikiMedia foundation. It is maintained by an active community that collaboratively edits the lexicon using the 'wiki' principles. Due to its broad scope it has become an important resource for NLP research (Zesch et al., 2008). Thus, there is a general interest in converting Wiktionary into a standard machine-readable form that can be directly exploited by NLP applications. As the pages in Wiktionary are actually very regularly structured, it is is in principle straightforward to extract the data. A Wiktionary page in particular consists of at least the following sections:

- A language block, containing all entries with the same orthographic form. For example the page *cat* contains the English word as well as the Indonesian word *cat* (meaning 'paint') and the Romanian word (meaning 'storey').
- Under each language block, the entries are then grouped by part of speech, i.e., the page for *bank* has both the noun and the verb listed together.
- Alternative forms.
- Pronunciations.
- The etymology.
- The body of each entry then consists of:
  - The inflectional information for the entry, e.g., "free (*comparative* freer, *superlative* freest)".
  - An enumerated list of definitions, often with usage notes such as "archaic" or "slang".
  - A list of synonym links.
  - A list of antonyms.
  - A list of derived terms.

– A list of translations.

We have developed a parser that works as a robust finite state automaton for parsing the XML dumps of Wiktionary. The automaton is illustrated in Fig. 3; it works for pages in English, German, French, Spanish, Dutch and Japanese.

Wiktionary:

```
<page>
<title>free</title>
==English==
===Adjective===
{{en-adj}}

# Not [[imprisoned]] or [[enslaved]].
# Obtainable without any [[payment]].

====Synonyms====
* {{sense|obtainable without payment}}:
    [[free of charge]], [[gratis]]

====Translations====
{{trans-top|not imprisoned}}
* German: {{t+|de|frei}}
{{trans-bot}}
</page>
```

lemon:

```
:free_en_adj lemon:canonicalForm [
  lemon:writtenRep "free"@en ] ;
  lexinfo:partOfSpeech lexinfo:adjective ;
  lemon:sense :free_en_adj_sense0 ;
  lemon:sense :free_en_adj_sense1 ;
  lemon:sense :free_en_sense_def .

:free_en_adj_sense0 lemon:definition [
  lemon:value "Not imprisoned or enslaved"@en ] ;
  lemon:reference
    <http://en.wiktionary.org/wiki/free> ;
  lexinfo:translation :frei_de_sense_def .

:free_en_adj_sense1 lemon:definition [
  lemon:value "Obtainable without any payment"@en ] ;
  lemon:reference
    <http://en.wiktionary.org/wiki/free> ;
  lexinfo:synonym :free_of_charge_en_sense_def .
```

Fig. 2 An example of a Wiktionary entry and the corresponding *lemon* generated

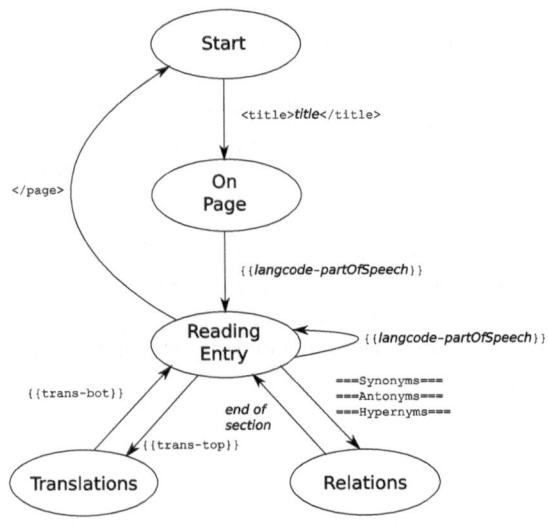

Fig. 3 The algorithm for extracting information from Wiktionary pages

Before a *lemon* model can be created for Wiktionary, a major issue is the definition of appropriate senses. As Wiktionary lists a number of definitions of the term, one could assume that they could be directly mapped to concepts in *lemon*. However, as there are different definitions per section of the article and there is no direct correspondence between these definitions, the task of collapsing various senses into an appropriate subset is not trivial. In a first step, we thus create one *lemon* sense for each definition. Then, we attempt to align the different definitions by computing the Levenshtein distance between the definitions in various sections. In the sections synonym, antonym, derived forms and translation (henceforth "SADT"), each sense used there has a gloss; *gratis*, for example, is specified as a synonym of *free* with a gloss 'obtainable without payment', and we assume that this corresponds to a definition given in the main definition section. We have observed that the ordering of definitions is similar to that of SADT senses and glosses are often short substrings of SADT glosses. Thus our algorithm for finding alignments, given a threshold λ, is:

- For each SADT sense s:
  - For each main sense m that is not already equated to some SADT sense:
    - If s's gloss is a substring of m's gloss, equate s and m; go to next sense;
    - Else, calculate the normalized Levenshtein distance between the glosses of s and m.
  - Select that main sense m that minimizes the Levensthein distance to s. If the corresponding Levenshtein distance is lower than λ, equate s and m.

The evaluation results of this algorithms that collapses senses together are presented in Tab. 1 in terms of "coverage" (indicating the percentage of senses that were mapped to a sense from the definition section), and "precision" (indicating the correctness of the mappings based on a sample of 100 randomly selected examples at each threshold level). The precision is indicated for various thresholds of the Levenshtein distance. Precision obviously increases with higher values for the threshold, but never drops below 71%. This is due to the fact that many entries have only one sense in the main definition, such that there is only one mapping candidate for the senses in the sections corresponding to SADT words.

Table 1 The results of merging duplicate senses found within Wiktionary.

	Merged	Coverage	Precision
Substring	36 595	37.8%	99.5%
> 0.9	6 842	44.9%	100%
> 0.8	3 398	48.4%	99%
> 0.7	2 669	51.2%	99%
> 0.6	3 243	54.5%	97%
> 0.5	7 128	61.9%	97%
> 0.4	4 612	66.6%	98%
> 0.3	6 295	73.1%	91%
> 0.2	7 983	81.4%	92%
> 0.1	6 934	88.5%	73%
> 0.0	3 862	92.5%	71%

4.3 Linking

As there will be many lexical entries that are common to both resources, a further goal is to identify these duplicates and merge them. This is clearly not the same as finding synonyms or equivalent synsets/sense across resources. We apply an entry linking criterion for this purpose that was previously described in McCrae et al. (2011). This method proceeds by first finding entries that have the same canonical form in both resources (case was ignored). Then, we compare the part-of-speech tags of both lexical entries. Note that, as we have aligned both resources to Lex-Info (Cimiano et al., 2010), this amounts to a simple string comparison. If the tags differ, we infer that the entries are different. We then check whether the remaining properties of the entry are *similar* as follows: for each property p with value v, if the other entry has a different value for p, consider the entries as different. We then also check each (non-canonical) form of the entries; for each of these forms, we find those that on the other entry are *(property) similar*. We then reject the entry if there is such a similar form on the other entry with a different written representation. In Tab. 2, we present the results of linking in terms of the number of entries that were linked against those that were not linked.

Table 2 The results of linking Wiktionary to WordNet

	#Entries	Percent (WN)	Percent (Wikt)
Linked	63,478	21.0%	26.9%
Not Linked (Wiktionary)	172,674	–	73.1%
Not Linked (WordNet)	238,408	79.0%	–
Ambiguous	1,741	0.6%	0.7%

We found that the overlap WordNet and Wiktionary in terms of lexical entries amounts to roughly between 20% and 25%. To investigate the reason for the low overlap between Wiktionary and WordNet, we took 50 entries from Wiktionary and analysed the mapping to WordNet. Out of these, 28 were also contained in WordNet. Of the remaining 32, 9 were single words which were simply absent from Word-Net, e.g. *polysemic* or *abaciscus*. Further, 10 entries were compounds not present in WordNet, e.g. *false friend* and *apples and pears*. Two further entries were contained in Wordnet but with a different part-of-speech i.e. *raven* as an adjective and *to minute* as a verb. Finally, Wiktionary had a separate lexical entry for the plural noun *wares*,[5] while WordNet correctly only listed the singular form as a lexical entry. Thus, the resources seem largely complementary, a surprising result. Combining them might thus yield a lexicon that has significantly better coverage, and would therefore be of more use to applications that rely on machine readable dictio-

---

[5] In spite of having two entries for the plural and singular of *ware*, Wiktionary specified that *wares* is the plural of *ware*.

nary/lexica. This ultimately corroborate the usefulness of creating / linking lexica following the Linked Data principles.

5 Discussion and Conclusion

In this paper we presented a case study showcasing the publication and linking of lexical resources following the Linked Data principles. The conversion of WordNet to Linked Data was rather straightforward due to the fact that an existing RDF export was available. We have argued that when using an ad-hoc RDF format for publishing resources, changes to the RDF schema might become necessary if the underlying data structures change. This can be alleviated by using a principled model such as *lemon*.

The conversion of Wiktionary to Linked Data was more intricate as it represents a semi-structured resource that needs to be parsed appropriately before. We found that *lemon* seems an adequate model which revealed important flaws in the design of Wiktionary, i.e. the fact that correspondences between sense definitions in various sub-sections of the article are not explicitly modelled. To address this, we have proposed a simple yet effective algorithm to align the definitions across sections. Finally, we proposed an approach to linking lexical entries across WordNet and Wiktionary, showing that the overlap between the two resources was lower than expected. Integrating both resources promises to create a wide-coverage resource that can be exploited in NLP applications. Instead of creating a new lexical resource from these two resources, we have shown how we can create a virtual new resource by applying the Linked Data technologies, linking lexical entries across both resources. In our view, the adoption of Linked Data principles is thus a promising method for extending the life cycle of linguistic resources In order to integrate resources in a principled manner, a common model is needed. In this paper, we have proposed the *lemon* model for this and shown that it provides a principled model to which the lexical resources we have considered (WordNet and Witkionary) could be straightforwardly converted to.

Acknowledgements This work was developed in the context of the Monnet project, which is funded by the European Union FP7 program under grant number 248458, the CITEC excellence initiative funded by the DFG (Deutsche Forschungsgemeinschaft), and the Spanish national project BabeLData (TIN2010-17550).

References

Berners-Lee T (2009) Linked Data-The Story So Far. International Journal on Semantic Web and Information Systems 5(3):1–22
Chiarcos C (2010) Grounding an Ontology of Linguistic Annotations in the Data
  Category Registry. In: Proceedings of the 2010 International Conference on Lan-
  guage Resource and Evaluation (LREC)
Chiarcos C (this vol.) Interoperability of corpora and annotations. pp 161–179
Cimiano P, Buitelaar P, McCrae J, Sintek M (2010) Lexinfo: A declarative model
  for the lexicon-ontology interface. Web Semantics: Science, Services and Agents
  on the World Wide Web
Farrar S, Langendoen D (2003) Markup and the GOLD Ontology. In: Proceedings
  of Workshop on Digitizing and Annotating Text and Field Recordings
Fellbaum C (1998) WordNet: An electronic lexical database. MIT press Cambridge,
  MA
Kemps-Snijders M, Windhouwer M, Wittenburg P, Wright S (2008) ISOcat: Cor-
  ralling data categories in the wild. In: Proceedings of the 2008 International Con-
  ference on Language Resource and Evaluation (LREC)
Kipper-Schuler K (2005) Verbnet: A broad coverage, comprehensive verb lexicon.
  PhD thesis, University of Pennsylvania
Levin B (1993) English Verb Classes and Alternations: A Preliminary Investigation..
  University of Chicago Press, Chicago
McCrae J, Spohr D, Cimiano P (2011) Linking Lexical Resources and Ontologies on
  the Semantic Web with Lemon. The Semantic Web: Research and Applications
  245–259
McCrae J, Aguado-de Cea G, Buitelaar P, Cimiano P, Declerck T, Gomez-Perez A,
  Gracia J, Hollink L, Montiel-Ponsoda E, Spohr D, Wunner T (in press) Interchanging lexical resources on the semantic web. Language Resources and Evaluation
Montiel-Ponsoda E, Gracia J, Aguado de Cea G, Gómez-Pérez A (2011) Representing translations on the semantic web. In: Proceedings CW (ed) Proceedings of
  the 2nd International Workshop on the Multilingual Semantic Web 2011 (MSW
  2011), vol 775, pp 25–37
Zesch T, Müller C, Gurevych I (2008) Extracting lexical semantic knowledge from
  wikipedia and wiktionary. In: Proceedings of the Conference on Language Resources and Evaluation (LREC), Citeseer, pp 1646–1652

Integrating Lexical Resources Through an Aligned Lemma List

Axel Herold, Lothar Lemnitzer, and Alexander Geyken

Abstract This paper presents the modelling of a common meta-index for large modern and historical lexical resources of the DWDS project. Due to the different lexicographical principles and traditions employed for these resources as well as the different historical periods covered, such a meta-index cannot be modelled as a simple list of 1 : 1-correspondences between entries across different dictionaries. In order to model the occurring phenomena such as graphematic headword variance, homography, semantic change and differences in the semantic entry structure a more complex typed link structure is required.

1 Project Background

We report on one facet of a long-term project which aims at the integration of several lexical and textual resources in order to document the German language and its use at several stages. The integration affects the dictionaries which are procured at the project *Digitales Wörterbuch der deutschen Sprache* (DWDS),[1] a project of the *Zentrum Sprache* at the Berlin-Brandenburg Academy of Science and the Humanities (Klein and Geyken, 2010). The synchronization of various dictionaries via a common index enables the simultaneous querying and display of information which can be considered to be information about the same lexical entry. The possibility to formulate queries across its data sources – dictionaries, corpora, statistical tools – is the key feature of the lexical information system DWDS (see Klein, 2004 for a detailed discussion of the lexical information system).

Axel Herold · Lothar Lemnitzer · Alexander Geyken
Berlin-Brandenburgische Akademie der Wissenschaften, Jägerstraße 22/23, D-10117 Berlin
e-mail: {herold,lemnitzer,geyken}@bbaw.de

[1] http://www.dwds.de/

C. Chiarcos et al. (eds.), *Linked Data in Linguistics*,   35
DOI 10.1007/978-3-642-28249-2_4, © Springer-Verlag Berlin Heidelberg 2012

The DWDS comprises a broad range of lexical resources:

eWDG2   is a faithful digitized version of the *Wörterbuch der deutschen Gegen-*
wartssprache (WDG, 1962–1977) a six-volume printed dictionary compiled at
the *Akademie der Wissenschaften der DDR*. It contains 120 000 entries and is
represented in TEI-P5 compliant markup (Herold, 2011).

DWDSWB   is a new and continuously extended edition of the WDG which – in
addition to its WDG heritage – contains more contemporary German vocabulary,
i.e. lexical entries which have come into use after 1977. New entries for about
25 000 headwords will be compiled by the DWDS project team during the next
six years.

EtymWB   is a faithful digitized version of the three-volume *Etymologisches Wör-*
terbuch des Deutschen (Pfeifer, 1989), a printed dictionary compiled at the
Akademie der Wissenschaften der DDR in the 1980s. It contains entries for 8 000
morphologically simple main headwords and 14 000 derived minor headwords.

[1]DWB   is a faithful digitized version of the first edition of the *Deutsches Wörter-*
buch (DWB, 1854–1961). This 33-volume printed dictionary contains about
300 000 main headwords and a yet unknown number of related minor headwords. For a documentation of the editing process of [1]DWB see Dückert (1987);
Schmidt (2004) provides a helpful introduction to the (proper) use of this dictionary.

Besides these lexical resources, the DWDS provides:

- a time and genre balanced corpus of the German language of the 20th/21st century which comprises 100 million tokens for the period 1900–2000 and roughly
  ten million tokens for the first decade of the 21st century (Geyken, 2007);
- an extended, opportunistic corpus mostly compiled from electronic newspaper
  texts with a total of 2.7 billion tokens of which approximately one billion tokens
  are publicly accessible;
- statistical results of corpus analysis, e.g. word distribution patterns and a word-
  sketch like application (*Wortprofil*, see Geyken et al., 2009).

The corpus resources are searchable with DDC, a powerful linguistic search engine
(Sokirko, 2003). With the exception of [1]DWB, which will be published towards the
end of 2012, all data are searchable on-line at http://www.dwds.de/. Each
resource is displayed in its own panel, that accounts for the specific presentational
needs of the resource.

2 Linking Across Dictionaries

The DWDS project aims at providing explicit links across all its lexical resources on
the level of entries. Links between lexical resources and corpora are automatically
established on the basis of lemmas and are subject to changes when the linguistic

(pre)processing of the corpora is improved. In this paper we focus on the construction and maintenance of explicit linkage, i.e. on constructing and representing a list of equivalent lemmas.

2.1 The Building Blocks: Entries, Lemmas, Headwords

For the lexical resources the main access path is provided on the level of individual lexical entries. Each lexical entry E describes one or more lemmas L. Lemmas in turn are represented by one or more headwords H.

$$E := \{L_1, \ldots, L_n\} \; ; \; L := \{H_1, \ldots, H_m\} \tag{1}$$

Throughout this paper, headwords are noted exactly as they appear in the printed source, e.g. [1DWB]SONNTAG, [WDG]Sonntag 'Sunday'.

WDG and DWDSWB constrain their entries to contain a single lemma. In EtymWB entries etymologically related lemmas are grouped together and are typically described discursively. This discursive structure often prohibits splitting entries into smaller self-contained parts describing one lemma only. Due to the long editing period of [1]DWB there are many inconsistencies with regard to the structure of lexical entries. The majority of entries is self-contained and understandable without context regardless of their morphological complexity. However, there are instances of entry embedding, i.e. entries being located inside other entries where the exact location of the sub-entry cannot be ignored because it associates the sub-lemma with a distinct sense of the main lemma (e.g. see [1DWB]SONNTAG 'sunday' II 2 e). We follow a liberal approach to identifying entries in [1]DWB: They must contain at least one headword (not necessarily entry initial), they have to be self-contained and we try to minimize the size of entries, i.e. we try to maximize the number of entries.

2.2 Entry Equivalence

The primary goal of linking the lexical resources explicitly is to reach high accuracy in the matching of lexical entries of the involved resources. The user should see in the dictionary panels the information found in the selected dictionaries for exactly the same lemma. We therefore opted for a manually controlled automatic alignment process. The result is an aligned lemma list which contains the lemma equivalence relations across all lexical resources.

As long as only the 'modern' contemporary dictionaries DWDSWB, WDG and EtymWB are considered, a binary equivalence function is sufficient to express whether two lemmas are equivalent or not. Lemma and headword selection target the same user group and the dictionaries were created around the mid/end of the

20th century. Based on this notion of equivalence the EtymWB can easily be exploited as an etymological supplement to the DWDSWB and WDG. Most of the alignment between the three dictionaries could be achieved automatically based on headwords because the headword selection scheme is nearly identical.

There were some challenges, though, that required manual correction to differing degrees:

Homography/homonymy: Homonymy – or homography, its lexicographical incarnation – is a long-debated topic in lexicology and lexicography. Up to now, researchers only agree that there are various reasons to assign homonymy to a headword, the relevant reasons among them are (see Behrens, 2002 for a more complete overview):

1.   the lexical entries which are represented by a common headword have formal differences, e.g. in gender, inflection or conjugation (formal criterion);
2.   the lexical form has meanings which are completely unrelated, even from a diachronic point of view (semantic or etymological criterion).

These criteria are orthogonal and each of them is based on concepts which give room to interpretation. Consequently they lead to some variance in the lexicographic practice even in the case where one criterion is followed explicitly. For an overview of the lexicographical practice in various synchronic German dictionaries see Kempcke (2001). The WDG follows the formal criterion and applies it rather modestly (WDG, 1962–1977, p. 20 f.; Kempcke, 2001, p. 63 f.). While in the EtymWB's preface no mention is made of any criterion, it might go without saying for an etymological dictionary that the etymological criterion is applied and the same holds for [1]DWB. To give just one example: WDG establishes two lexical entries for [WDG]See, i.e. *der See* (masculine) 'lake', and *die See* (feminine) 'sea'. As there is no diachronic or etymological reason to tear these meanings apart, hence there is only one lexical entry in EtymWB and in [1]DWB respectively.

Since there is no single explicit criterion for ordering lexical entries with identical headwords, there is variance in the order of homograph entries between the dictionaries.

Incompatibility: By this term we mean that one dictionary supplies more homographs (entries) for a headword than the other. For example, EtymWB (and again, [1]DWB) presents a reading of [EtymWB]Art, roughly meaning 'ploughed land', which is not used in contemporary German and therefore missing in the WDG and DWDSWB. On the other hand, the WDG lists two homographs for [WDG]ober 'above' (as does [1]DWB), one being a variant which is only used in Austria, while EtymWB has only one entry with that headword.

Headword variance: According to some variance in selecting headwords that represent the lemma, we can find differing headwords for obviously identical lemmas across dictionaries. Often the cause for this is a regular choice among canonical forms, e.g. for de-adjectival nouns as in *(ein) Angesteller* vs. *(der) Angestellte* 'clerk' (indefinite vs. definite determiner). Another reason for headword variance lies in the strong tendency for the use of the plural form of a noun while the singular form is not completely ruled out and might even have been predominant in earlier language stages. This is the case with e.g. *Aliment* (singular) vs. *Alimente* (plural) 'alimony'.

Most of these issues presented obstacles to a straight-forward automatic alignment of entries which would have been based on the form of the headword alone.

We are currently working on the integration of [1]DWB, which poses some extra challenges on top of those which have been described above:

Homography/homonymy: Again, we face the consequences of different theories of homonymy across dictionaries leading to different numbers of homographs for a given lemma. Additionally, homographs in [1]DWB are not consistently marked and in some cases not marked at all (e.g. [1DWB]MAST with two entries for a male noun and two for a female noun). Currently there are more than 10 000 headwords known to appear multiple times in [1]DWB. As still more headwords are identified these numbers are even expected to increase.

Semantic change: The first volume of [1]DWB was issued in 1854 – more than 100 years before the first volume of WDG appeared in 1962. Consequently there is a considerable amount of lexical entries where the description of formally identical lemmas clearly shows radical semantic changes leeding to incongruent or even incompatible semantic descriptions. Let us consider a selection of lemmas that are recorded as non-homographic in [1]DWB and WDG:

- [1DWB]BARBAR vs. [WDG]Barbar – while [1]DWB describes the historical Greek meaning of the lemma only ('foreign people, foreign human, non-Greek'), we find three distinct sense descriptions in the WDG of which only the last one (explicitly marked as historical) corresponds with the sense description given in [1]DWB. The other two modern senses ('cruel human' and 'ignorant people, philistine') are derived from metaphorical uses of *Barbar* and are predominant in present-day German.

- [1DWB]GEBILDET vs. [WDG]gebildet – here, [1]DWB lists three senses for the lemma, namely 'illustrated', 'shaped, made of' and 'educated, intellectual' of which only sense number three is directly attested in the WDG. Following the pointer to the derivational base *bilden* which is also given in the WDG, the second sense can be deduced by the reader even though it is not stated explicitly. However, the first sense attested in [1]DWB remains exclusive to this dictionary.

- An example for a complete meaning shift is [1DWB]TRILLION vs. [WDG]Trillion where the former describes it as *tausend billionen* (10^{15}) and the latter describes it as 10^{18} Unfortunately there is no semantic paraphrase for *billion* in [1]DWB and some other powers of ten like *billiarde* do not appear as headwords at all. This phenomenon of incompatible meanings is also frequently found in the semantic description of man-made artifacts like [1DWB]HOLZSTOCK 'chopping block' and 'printing block' vs. [WDG]Holzstock 'wooden stick'.

Under- or even unregulated orthography: Special forms of headword variance are due to the hazards of an under-regulated, if not unregulated orthography at least in the first half of the time in which [1]DWB was compiled.

In cases where orthographical changes lead to a headword appearing under another initial letter again, "continuation entries" were inserted effectively resulting in two entries describing exactly the same lemma. The continuation typically consists of complementary information. Consider for example [1]DWB CAPELLE 'chapel'. This entry appeared in an 1855 partial issue, before (in an 1864 partial issue) [1]DWB KAPELLE 'chapel' was introduced pointing to the older entry and extending it with a five-fold explicit sense description and many more usage examples.

In [1]DWB, there are even occurrences of headwords that are judged as orthographic "errors" by the editors which didn't stop them from writing an article about that lemma, e.g. [1]DWB CAPELLE *fehlerhaft für cupelle, cupella* 'wrongly (written) for cupelle, cupella'.

The rather late fixing of orthographical norms – the first official norm for written German was decided in 1902 –, as well as various changes of this norm since then, cause differences in the form of headwords which are hard to capture. One typical pattern is the letter sequence *th* that was replaced by *t* in many (but not all) words, e.g. *Thal* $\longrightarrow$ *Tal* 'valley'. Another frequent pattern is the variance between *c* and *k*. There are regular patterns of orthographic change which we will try to capture with an orthographic normalizer. We use CAB (cascaded analysis broker) for this purpose, a rule based transducer which originally maps historical to contemporary spelling (Jurish, 2010). The technology, however, allows us to use the program the other way round and to produce possible historical spelling variants of contemporary headwords. The overgeneration of forms can be controlled by matching all output forms with the [1]DWB headwords list.

However, there are still many so-called "false friends" which are an obstacle to any automatic alignment. Let us illustrate this point with an example: Naïve alignment would run into problems with headwords like DWDSWB Turm. There is a corresponding headword in [1]DWB, but this one represents another lemma (which is not in use in contemporary German). The correct correspondence is [1]DWB THURM.

Idiosyncratic canonicalization of headwords: For reasons which are hardly understandable nowadays, the lexicographers chose to use canonical forms as headwords which look quite idiosyncratic to the contemporary user and lexicographer. First, all headwords in [1]DWB appear completely in capital letters, while in DWDSWB as in most other contemporary German dictionaries nouns start with a capital letter while words of other parts of speech do not, e.g. DWDSWB Weg vs. [1]DWB WEG 'way' and DWDSWB weg vs. [1]DWB WEG 'gone'. For a correct mapping of these entries, the part of speech information given in both dictionaries can be used, e.g. *masculine noun* vs. *adverb* for [1]DWB WEG. To complicate matters, though, part of speech information is not available for all entries in [1]DWB.

The letter *ß* (eszett or sz-ligature) is specific to the German writing system. It is represented literally as *SZ* in [1]DWB headwords, e.g. [DWDSWB]Spaß vs. [1]DWB SPASZ 'fun, joke'. This is effectively a lossy transformation and cannot be simply reversed without morphological analysis because the letter sequence *sz* is of course completely legal in German, e.g. [DWDSWB]Lebenszeichen 'vital sign'.

Historical and dialectal lemmas: Due to the dictionary's decidedly diachronic view there is a considerable amount of lemmas that were in use in historic times but are not attested in DWDSWB or EtymWB because they fell out of general use. Typically those lemmas cannot be found in present day corpora except, perhaps, as common names, e.g. [1]DWB SESTER, a historical measure of capacity. Another group of lemmas that are not found in today's general dictionaries are only used in certain dialects, many of which are not explicitly marked as such in [1]DWB, e.g. Lower German [1]DWB PADDEN 'making small steps (like a toad)'.

Although formal variance in headwords representing the same lemma is an obstacle for automatic matching of entries across dictionaries this phenomenon can be accounted for by acknowledging different lemmatisation strategies and allowing for different orthographies. However, due to different accounts to homography and more so incongruent or even incompatible semantic descriptions an aligned lemma list cannot be modelled as a simple list of 1 : 1 correspondences between lexical entries.

3 Evaluation

DWDSWB, WDG, and EtymWB are completely aligned and the alignment is actively exploited on the project's website to achieve a synchronized display of equivalent lexical entries. This makes it possible to use EtymWB as an etymological extension to the synchronous view of the present-day dictionaries.

First alignment tests between DWDSWB and [1]DWB on a random sample (941 entries) of approximately 45 000 non-homographic entries appearing in both dictionaries show strong semantic equivalence for about 67 % (632) of those entries. Another 3 % (27) are clearly not semanticly related, i.e. their semantic descriptions are incompatible. For about 8 % (79) we found partial semantic overlap. These pairings of entries are instances of incongruent semantic descriptions. The remaining entries 22 % (203) cannot be evaluated based on the entries alone, typically because DWDSWB lists a considerable amount of lemmas only with their headword(s) and still lacks semantic descriptions for them. During the course of the DWDS project these omissions will be amended.

While the figures for headword identical entries already look promising there is still much space for improving the alignment strategy considering the total number of entries and headwords in both dictionaries. To achieve a wider coverage on the non-homographic entries we will use CAB as a filter to derive additional mappings between contemporary and historical headwords. These suggestions have of course to be accepted or dismissed by a skilled lexicographer. Our aim is still to cover as many entries of these resources as possible.

4 Reusable Representation

As a result of our alignment efforts we want to provide to the community a combined lemma list of our lexical resources which will encompass synchronic as well as diachronic dictionaries and which will enable other lexicographic data centres to align their resources with this list. There are different technical standards to represent an aligned lemma list in a reusable way:

RDF (Resource description framework):   RDF is very generic in allowing to express relations among arbitrary entities. Currently we are working towards a taxonomy of entry relations based on the phenomena sketched in Sect. 2 that will allow to express the relations among the dictionaries in RDF statements.

LMF (Lexical markup framework):   LMF is an emerging standard directed specifically towards the representation of lexical resources. There are two basic modelling options with regard to LMF:

- transferring DWDSWB into LMF and providing links to the other dictionaries (e.g. via LMF's multilingual notations extension) which in turn would not have to be modelled in LMF;
- considering the aligned lemma list as a resource in its own right and model it in LMF, again providing links to the dictionaries.

The second option allows for easier maintenance of headword variants and can be directly exploited for queries across the represented dictionaries. It allows linkage among arbitrary dictionaries because it does not require a 'central' dictionary to contain a semantically equivalent entry. Most importantly, it provides a clear distinction between the underlying lexical resources and the relations among them similar to the RDF representation.

The aligned lemma list will be made available in RDF and LMF.

5 Future Work

We clearly need a solid operationalisation for the concept of semantic equivalence that allows for robust classification. Given the huge amount of manual effort needed to complete the alignment between DWDSWB and [1]DWB on the level of lexical entries it seems unfeasible to achieve a mapping for individual senses.

The lexical entries of DWDSWB, WDG, EtymWB and [1]DWB will become publicly exposed through the project CLARIN (Common Language Resources and

Technology Infrastructure) where persistent identifiers will be employed to provide a stable reference system across different versions of the dictionaries.

In the near future we are also going to extend the aligned lemma list with headwords from GermaNet (Kunze and Lemnitzer, 2010). Since GermaNet is a resource which orders headwords according to individual senses, we expect further challenges to the alignment.

Acknowledgements This work was supported by the long term project DWDS of the Berlin-Brandenburg Academy of Science and the Humanities and CLARIN-D (Common Language Resources and Technology Infrastructure, `http://de.clarin.eu/`), funded by the German Federal Ministry for Education and Research.

References

Behrens L (2002) Structuring of word meaning II: Aspects of polysemy. In: Cruse DA, Hundsnurscher F, Job M, Lutzeier PR (eds) Lexikologie – Lexicology. Ein internationales Handbuch zur Natur und Struktur von Wörtern und Wortschätzen, vol 1, de Gruyter, Berlin, pp 319–337

DWB (1854–1961) Deutsches Wörterbuch. Hirzel, Leipzig

Dückert J (ed) (1987) Das Grimmsche Wörterbuch. Untersuchungen zur lexikographischen Methodologie. Hirzel, Leipzig

Geyken A (2007) The DWDS corpus: A reference corpus for the German language of the twentieth century. In: Fellbaum C (ed) Idioms and collocations: Corpus-based linguistic and lexicographic studies, Research in corpus and discourse, Continuum, London, pp 23–40

Geyken A, Didakowski J, Siebert A (2009) Generation of word profiles for large German corpora. In: Kawaguchi Y, Minegishi M, Durand J (eds) Corpus analysis and variation in linguistics, Studies in Linguistics, vol 1, John Benjamins, pp 141–157

Herold A (2011) Retrodigitalisierung und Modellierung des Wörterbuchs der deutschen Gegenwartssprache. In: Krafft A, Spiegel C (eds) Sprachliche Förderung und Weiterbildung – transdisziplinär, no. 51 in Forum Angewandte Linguistik, Peter Lang, Frankfurt (M.), Berlin

Jurish B (2010) More than words. Using token context to improve canonicalization of historical German. JLCL 25(1):23–40

Kempcke G (2001) Polysemie oder Homonymie? Zur Praxis der Bedeutungsgliederung in den Wörterbuchartikeln synchronischer einsprachiger Wörterbücher der Deutschen Sprache. Lexicographica 17:61–68

Klein W (2004) Vom Wörterbuch zum Digitalen Lexikalischen System. Zeitschrift für Literaturwissenschaft und Linguistik 136:10–55

Klein W, Geyken A (2010) Das digitale Wörterbuch der deutschen Sprache (DWDS). Lexicographica 26:79–93

Kunze C, Lemnitzer L (2010) Lexical-semantic and conceptual relations in Ger-
    maNet. In: Storjohann P (ed) Lexical-semantic relations: Theoretical and practi-
    cal perspectives, no. 28 in Lingvisticæ Investigationes Supplementa, John Ben-
    jamins, Amsterdam, pp 163–183
Pfeifer W (1989) Etymologisches Wörterbuch des Deutschen. Akademie-Verlag,
    Berlin
Schmidt H (2004) Das Deutsche Wörterbuch. Gebrauchsanweisung. In: Bartz HW,
    Burch T, Christmann R, Gärtner K, Hildenbrandt V, Schares T, Wegge K (eds)
    Deutsches Wörterbuch. Elektronische Ausgabe der Erstbearbeitung von Jacob
    Grimm und Wilhelm Grimm, Zweitausendeins, Frankfurt (M.), pp 25–64
Sokirko A (2003) DDC – a search engine for linguistically annotated corpora. In:
    Proceedings of Dialog 2003, Protvino (Russia)
WDG (1962–1977) Wörterbuch der deutschen Gegenwartssprache. Akademie-
    Verlag, Berlin

Linking Localisation and Language Resources

David Lewis, Alexander O'Connor, Sebastien Molines, Leroy Finn,
Dominic Jones, Stephen Curran, and Séamus Lawless

Abstract Industrial localisation is changing from the periodic translation of large bodies of content to a long-tail of small, heterogeneous translations processed in an agile and demand-driven manner. Software localisation and crowd-source translation already practice continuous fine-grained distribution of translation work. This requires close integration and round-trip interoperability between content creation and localisation processes, while at the same time recording the provenance of translated content to maximise it reuse in future translation tasks, and, increasingly, in training Statistical Machine Translation (SMT) engines. This work adopts a Linked Data approach to integrating the content translation round-trip process with the logging of process quality assurance provenance. This integration supports a pull-based interoperability model that supports continuous synchronising of content and process meta-data between the generating organisation and any number of language service providers or translators. We present a platform architecture for sharing, searching and interlinking of Linked Localisation and Language Data (termed L3Data) on the web. This is accomplished using a semantic schema for L3Data that is compatible with existing localisation data exchange standards and can be used to support the round-trip sharing of language resources. The paper describes our approach to development of L3Data schema and data management processes, web-based tools and data sharing infrastructure that use it. An initial proof of concept prototype is presented which implements a web application that segments and machine translates content for crowd-sourced post-editing and rating.

David Lewis · Alexander O'Connor · Sebastien Molines · Leroy Finn · Dominic Jones ·
Stephen Curran · Séamus Lawless
Centre for Next Generation Localisation, Knowledge and Data Engineering Group, Trinity College Dublin, Ireland e-mail: {Dave.Lewis,Alex.OConnor,moliness,finnle,
Dominic.Jones,Stephen.Curran,slawless}@cs.tcd.ie

C. Chiarcos et al. (eds.), *Linked Data in Linguistics*,   45
DOI 10.1007/978-3-642-28249-2_5, © Springer-Verlag Berlin Heidelberg 2012

1 Introduction

Industrial localisation is changing from the periodic translation of large bodies of content to a long-tail of small, heterogeneous translations processed in an agile and demand-driven manner. To enable service providers to compete and innovate in this new market, the capture, reuse and sharing of multilingual data generated in the localisation process becomes essential.

This data is used to train high-quality, domain-specific data-driven Language Technology services . These services include Statistical Machine Translation (SMT) and text classifiers. This training process is key to addressing long-tail localisation efficiently..

This work adopts a Linked Data approach (Bizer et al., 2009) to the massively scalable sharing, searching and interlinking of Localisation and Language Linked Data (termed here L3Data) on the web. This is accomplished using a semantic schema for L3Data that is compatible with existing localisation data exchange standards and can be used to support the round-trip sharing of language resources. Multilingual Terminology Databases, or Term-Bases, and Translation Memories are shared as they have the greatest impact on conventional localisation productivity and on the training of relevant language technologies such as SMT. Commercially sustainable sharing of L3Data must work within constraints of copyright, confidentiality and competitive concerns (Allemang, 2010), but by reflecting these in the L3Data schema, fine-grained access control rules allow enterprises to flexibly balance them against the benefits of sharing data.

The paper describes our approach to development of L3Data schema and data management processes, web-based tools and data sharing infrastructure that use it. An initial proof of concept prototype is presented which implements a simple web service for segmenting and machine translating content for crowd-sourced post-editing and rating, built on a triple store that tracks steps in this workflow using an integration of the Open Provenance Model (Moreau et al., 2008) and the XLIFF localization interchange format from OASIS (XLIFF, OASIS, 2007).

2 Motivation

In the software industry, the trends towards software as a service offering and smartphone apps has greatly levelled the playing field for small and medium enterprises (SMEs) entering a global market. At the same time, technical documentation has shifted from monolithic technical documents to forms such as FAQ, knowledge articles, wikis and Question-Answer forums, which contain high proportion of user-generated content or vendor content produced by customer care staff asynchronously from the product release (Marcus, 2006). For the localisation industry this represents an increasing challenge in terms of growing demand for content translation coupled with reduced job size and increased heterogeneity of content style and domain, which can be characterised as long tail localisation. However, in recent decades the principle efficiency gains in the localisation industry have resulted in the sharing of Translation Memories (TM, enabling maximal reuse of previous translations) and Term-Bases (TB) between client, localisation service providers (LSPs) and translators. These resources amount to multilingual terminology databases that ensure consistent use of terms in the source content and consistent translation of those tools. However, the characteristics of long tail localisation tend to weaken these efficiency gains, and especially for SMEs in the localisation chain, reduce the attractiveness of investing in the assembly and maintenance of TMs and TBs.

3 Approach

In summary, SMEs now have near-transparent global distribution opportunities, but they are faced with the need to support a bewildering number of locales and languages to gain the benefit of such worlwide availability. To address this localisation barrier for SMEs engaged as provider or consumers in long tail localisation, we propose a solution that combines the carefully targeted and controlled sharing of language resources with open innovation in the use of data-driven language technology services. Specifically this project will develop controlled sharing solutions for multilingual terminology management to support consistency in source language content authoring and its translation and for parallel text management for leverage as TMs and for the training of statistical machine translation (SMT) engines. The premise is that effective use of language resources in long-tail localisation requires low-cost acquisition of domain-specific, quality-assured resources that are best sourced from prior localisation tasks provided these have systematically integrated quality-related provenance annotation into the resources they produce. The sharing solution takes the form of Linked Data on the Web. This builds on the W3C's Semantic Web Standards RDF, RDFa, RDF(S) and OWL, which allows web content and web data (i.e. deep web content) to be interlinked in a decentralised and distributed manner. The approach is inherently multi-lingual as Linked Data supports multi-lingual data and meta-data representation enabled through Unicode, element-level language tagging in RDF and International Web Resource identifiers.

A key requirement is therefore to develop an extensible meta-data schema for localisation data to support fine-grained, low cost interlinking of source language web content and language resources in the form of multilingual term-bases and TMs as used in the localisation process. This schema will also enable terminology and translations generated in the localisation process to be harvested for further sharing, including its use for SMT training. We refer to the subject of this Linked Data schema as Linked Localisation and Language Data or L3Data. This schema will then support the development of terminology management, parallel corpora management and translation management tools that establish and maintain the links between localised content, its localisation-related meta-data and the shared language resources leveraged during the localisation process. Where appropriate these will be developed as plug-ins to existing localisation support products. Our aim is specifically not therefore to establish an open access repository of Localisation and Language Linked Data, but to provide instead an open data-linking platform that, with suitably fine-grained and auditable access control, will support low-cost data sharing suitable for new commercial value networks that address long tail localisation. This platform will also then support new innovative localisation support services such as: rapid, domain specific SMT training; interlink discovery services to integrate L3Data in support of new domain-specialised value chains and data cleaning and link maintenance services.

3.1 Support for SMT Training

Data-driven language technologies are demonstrating benefits at different stages of the localisation process such as: SMT to support human translation, named entity recognition for terminology management and text classification for translation review. The development of bespoke SMT engines in particular has become increasingly accessible to SMEs thanks to the popularity of the MOSES open source toolkit (Koehn et al., 2007). However, the effectiveness of these technologies is highly sensitive to the relationship of the data upon which they have been trained to the localisation task at hand. Assuring this in long tail localisation requires low cost, but highly agile and responsive acquisition of training data. This can be achieved by continuous and targeted sharing of linguistic resources that are the by-products of localisation – i.e. translation memories, term-bases and question-answer (QA) annotations of these – across the broadest range of LSPs and their clients. Long-tail localisation inherently requires data linking as the fragmentary nature of the content means that the heterogeneous data must be collated and links kept up-to-date, and places even greater value on agility in sharing to be able to assemble sufficient data to support rapidly changing application domains. Until now there has not been an effective model for rapid language data exchange with suitable provenance needed to address commercial quality and access control concerns.

3.2 Relationship to Existing Localisation Standards

However, this requires high levels of interoperability, both in the linguistic resources that are shared and the language technology-based localisation support services, often from third parties, that would operate over them. Language resources, even when made openly available, exist in information silos with their own distinct schema and access portals. Even when standardised exchange formats are available, such as TermBase Exchange (TBX) for terminology (Localization Industry Standards Association, 2008), Translation Memory Exchange (TMX, Localisation Industry Standards Association, 2005) or XLIFF for localisation job hand-off format (XLIFF,

OASIS, 2007), they are implemented through tool import/export functions that do not support round-trip consistency management as the language resources are updated over time. Though there have been some proposals to export existing language resources in RDF, such as an RDF mapping for ISO TC37 data categories (see, e.g., Windhouwer and Wright, this vol.), these efforts focus more on rich linguistic annotation of semantic web ontologies (Buitelaar et al., 2009) rather than commercial terminology management for which TBX provide a suitable lightweight starting point. Based on a subset of ISO TC37 data categories, the potential for extending with these richer lexical schema remain. The Multi Lingual Information Framework (Cruz-Lara et al., 2005) extends the ISO data categories work to integrate localisation meta-data as exchanged in TMX and XLIFF, but has remained at the specification stage and does not attempt to exploit Linked Data. The OASIS OAXAL initiative (Zydroń, 2011) proposed mechanisms for linking from source web content to term-base entries (using the Termlink XML schema) and segment-level translations (using the XML Text Memory, xml:tm schema), and can therefore be easily extended to reference L3data potentially held by partner organisations.

3.3 Use of Linked Data Infrastructure

The L3Data approach combines Linked Data architectures with fine-grained access control to address the need of data to be distributed but connectable in the rapidly forming value chains. These are the key properties needed to address the heterogeneous but short-lived jobs typical of long tail localisation. The dereferencable property of RDF-based Linked Data allows decentralised development of unambiguous schema terms that can be subsequently interlinked, potentially by third parties. The approach enables rapid specialisation of language resources to support innovation in what data and meta-data can be readily shared. The W3C's standardisation of RDF and RDFS to author schemas and publish data directly on the web and RDFa to link XML content to RDF elements has resulted in a growing range of performant and robust Linked Data stores ('triple stores') with web accessible query interfaces (for which a standard query language SPARQL, is typically used). However, localisation support systems typically place emphasis on data handling performance, especially for interactive translation tasks such as TM searches. However, an initial comparison on TM look up, using a 3 million word TM typical of the size supported by desktop translation tools, showed only a 20% overhead incurred by an unoptimised off-the-shelf triple store compared to the leading tool (SDL's Trados™).[1]

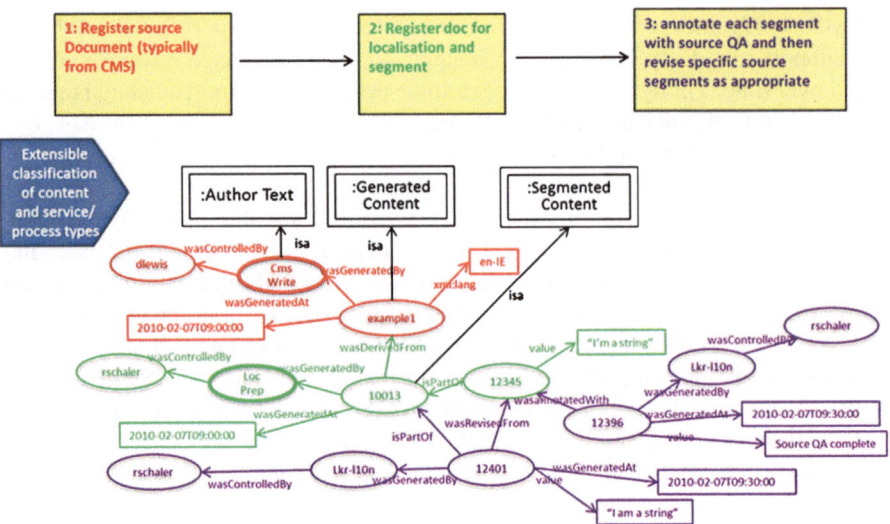

Fig. 1 Example of RDF provenance capture for a three step workflow

4 Implementation

Despite working in a domain that requires controlled access to Linked Data, we follow the decentralised evolution principles of Linked Open Data, in making use of
and extending existing RDF vocabulary, specifically the Open Provenance Model
(Moreau et al., 2008, OPM). A content processing approach is therefore taken for
the L3Data schema, where OPM is used to express the state transformation that
operates on content (principally terms and translation units) and its meta-data as
the result of content processing by different processes such as authoring/revision,
source language question-answering (including terminology usage), TM leverage,
SMT usage, human translation, post-editing, target language question-answering
etc. Figure 1 shows an example of the RDF provenance data captured for three steps
of registering a source document with the system, preparing it for localisation by
segmenting and running a controlled language quality check on those segments.
This approach can also be used to record value adding operation to shared TM or
TB elements, such as adding term translations, definitions or morphologies or identifying terms, or style and domain classification TM entries. OPM serves to capture
both the process resulting in the recorded content transform or annotation and the
agent responsible, thereby support the management of acknowledgement and credit
for shared language resources.

Our initial implementation focused on recording transforms produced by external service in a translation workflow that used XLIFF to exchange job data between
different web services implemented as part of the Service Oriented Localisation

---

[1] http://www.trados.com

Fig. 2 Provenance visualiser screenshot

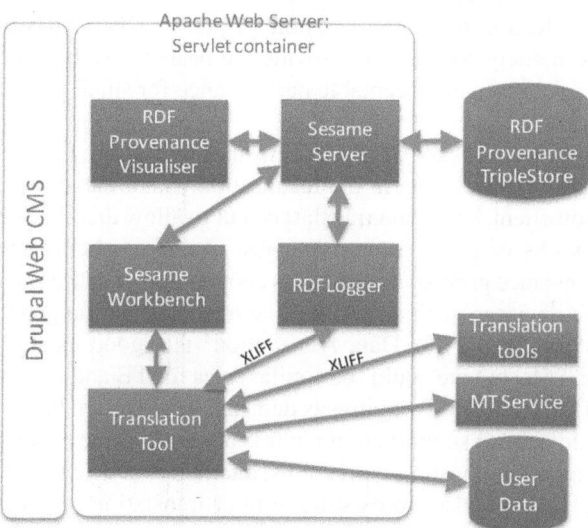

Fig. 3 Content Management System (CMS) Localisation System architecture

Architecture Solution (SOLAS) platform developed by the Localisation Research Centre at the University of Limerick, Ireland. A generic XLIFF to RDF/OPM transformer was implemented, while user defined extensions in the XLIFF recording the workflow activity routing was used to populate the process property of the activity transform. The different forms of translated content and the processes performed on them were classified from taxonomies for localisation content and service categories (Lewis et al., 2010) defined by the Centre for Next Generation Localisation (van Genabith, 2009).

To demonstrate this approach a crowdsourced translation application which has been implemented with a Drupal frontend, via which users can create and contribute to translation jobs in XLIFF. An RDFLogger component is used to change the XLIFF document into RDF provenance statements and then logs these to a triple store. Further, the Sesame Triple Store[2] provides an open source Java framework for storing, querying and reasoning with RDF. A RDF Provenance Visualiser (Fig. 2) has been implemented for exploring outcomes of process steps using the provenance based logging of localisation activities. Subsequently a second application offering a simple web based crowd-sourced translation and rating application was developed and is currently being evaluated. Figure 3 outlines the implementation architecture.

5 Future Directions

Our initial evaluation and proof of concept implementation show that a provenance based Linked Data approach to localisation interoperability is possible, can interoperate with existing standards, namely XLIFF, can provide the basis for web-based translation applications, and operates with acceptable performance for small jobs at least.

Our next steps will be working to establish a platform that will allow L3Data to be shared between the conventional actors in the localisation chain, namely the content developing localisation client, LSPs and translators, but to allow this sharing to occur between value networks of peers (rather than just chains of clients and providers (Allee, 2002)), for instance groups of translators cooperating in addressing a large customer domain. We envisage such shared will be mediated by Language Data Resource (LDR) Curators. The TAUS Data Association[3] is a good existing example of such a curator for TMs, but we could see similar roles for TB and for the interlinking and assembly of LDR for SMT training as depicted in the figure below.

The L3Data platform would need to support common features such as the posting of LDR to a value network (including posting corrections, alternates, new language translation or annotation such as 'see also' to existing term or translation segment resources), interlink management, access control (possibly using content annotation using RDF encodings of Creative Commons), annotating LDR with question-

---

[2] http://www.openrdf.org
[3] http://www.tausdata.org

answering outcomes from translation work (e.g. a poor question-answering rating on reuse of a previously submitted TM segment in a particular context) and auditing of LDR usage in a job conducted within an actor, (in order to help gauge the benefits yielded by participating in a LDR sharing value network)

Acknowledgements This research is supported by the Science Foundation Ireland (Grant 07/CE/I1142) as part of the Centre for Next Generation Localisation (http://www.cngl.ie) at Trinity College Dublin.

References

Allee V (2002) The Future of Knowledge: Increasing Prosperity through Value Networks. Butterworth-Heinemann

Allemang D (2010) Semantic web and the linked data enterprise. In: Woods D (ed) Linking enterprise data, Springer, pp 3–23

Bizer C, Heath T, Berners-Lee T (2009) Linked data - the story so far. International Journal on Semantic Web and Information Systems 5:1–22

Buitelaar P, Cimiano P, Haase P, Sintek M (2009) Towards linguistically grounded ontologies. In: Proceedings of the 6th European Semantic Web Conference (ESWC 2009), Heraklion, Greece, LNCS, vol 5554, pp 111–125

Cruz-Lara S, Gupta S, García J, Romary L (2005) Multilingual information framework for handling textual data in digital media. In: Proceedings of the 3rd International Conference on Active Media Technology (AMT 2005), Kagawa, Japan, pp 81–84

van Genabith J (2009) Next generation localisation. Localisation Focus: The International Journal of Localisation 8:4–10

Koehn P, Hoang H, Birch A, Callison-Burch C, Federico M, Bertoldi N, Cowan B, Shen W, Moran C, Zens R, Dyer C, Bojar O, Constantin A, Herbst E (2007) MOSES: Open source toolkit for statistical machine translation. In: Proceedings of the 45th Annual Meeting of the Association for Computational Linguistics (ACL 2007). Companion Volume Proceedings of the Demo and Poster Sessions, Prague, Czech Republic, pp 177–180

Lewis D, Curran S, Jones D, Moran J, Feeney K (2010) An open service framework for next generation localisation. In: LREC 2010 Workshop on Web Services and Processing Pipelines in HLT: Tool Evaluation, LR Production and Validation, Valetta, Malta, pp 52–59

Localisation Industry Standards Association (2005) TMX 1.4b Specification OSCAR Recommendation. http://www.lisa.org/fileadmin/standards/tmx1.4/tmx.htm, retrieved on 25 Feb 2010

Localization Industry Standards Association (2008) Systems to manage terminology, knowledge, and content – TermBase eXchange (TBX). http://www.lisa.org/TBX-Specification.33.0.html, retrieved on 25 Feb 2010

Marcus A (2006) A demand-based view of support: From the funnel to the cloud. Tech. rep., Service Innovation Consortium, San Carlos, CA, retrieved 18/8/11

Moreau L, Freire J, Futrelle J, McGrath R, Myers J, Paulson P (2008) The open provenance model: An overview. In: Freire J, Koop D, Moreau L (eds) Provenance and Annotation of Data and Processes, LNCS, vol 5272, Springer Berlin / Heidelberg, pp 323–326

Windhouwer M, Wright SE (this vol.) Linking to linguistic data categories in ISOcat. pp 99–107

XLIFF, OASIS (2007) Xliff 1.2. a white paper on version 1.2 of the xml localisation interchange file format (xliff). http://xml.coverpages.org/XLIFF-Core-WhitePaper200710-CSv12.pdf, revision: 1.0, 17 Oct, retrieved on 25 Feb 2010

Zydroń A (2011) Reference model for open architecture for XML authoring and localization 1.0 OASIS committee specification. http://www.oasis-open.org/committees/oaxal/, retrieved 18/8/11

Part II
Corpus Building and Annotation

Reusing Linguistic Resources: Tasks and Goals for a Linked Data Approach

Marieke van Erp

Abstract There is a need to share linguistic resources, but reuse is impaired by a number of constraints including lack of common formats, differences in conceptual notions, and unsystematic metadata. In this contribution, the five most important constraints and the tasks necessary to overcome these issues are detailed. These constraints lie in the design of linguistic resources, the way they are marked up and their metadata. These issues have also come up in a domain other than linguistics, namely in the semantic web, where the Linked Data approach proved useful. Experiences and lessons learnt from that domain are discussed in the light of standardisation and reconciliation of concepts and representations of linguistic annotations.

1 Introduction

Linguistic resources, which form the core of Natural Language Processing (NLP) research and applications, are expensive to generate. Most research uses some sort of manually annotated data which requires extensive human effort to create. The best corpora involve hundreds (if not thousands) of man-hours. Furthermore, even the results generated from the application of automated techniques still involve complex tuning and parametrisation of algorithms, making them resources not easily reproduced. Because of the expense in producing linguistic resources, the NLP community often shares and reuses its datasets. However, the reuse of linguistic resources is not straightforward and limited because of a number of constraints. In this contribution, these constraints are specified, their influence on reusability is explained and a path towards a solution is discussed that draws upon lessons learnt from the Link-

Marieke van Erp
Web and Media Group, Computer Sciences Department, VU University, De Boelelaan 1081a, 1081 HV Amsterdam, The Netherlands e-mail: `marieke@cs.vu.nl`

C. Chiarcos et al. (eds.), *Linked Data in Linguistics*,
DOI 10.1007/978-3-642-28249-2_6, © Springer-Verlag Berlin Heidelberg 2012

ing Open Data Project.[1] At the time of writing this contribution, the LOD2[2] project
publicly announced that they are developing an NLP Interchange Format (NIF)[3]
to further interoperability between NLP tools, linguistic resources and annotations
(Hellmann et al., this vol.). Interoperability is also addressed several contributions
in this volume, including Chiarcos (this vol.), Eckart et al. (this vol.), Windhouwer
and Wright (this vol.), and it is one of the proclaimed goals of the Open Linguistics Working Group (Chiarcos et al., this vol., see there for related activities beyond
the Semantic Web community). The emergence of such initiatives from the Linked
Data community indicates that there is a desire from this community to use and
reuse linguistic resources.
The following five constraints are discerned:

1. Linguistic resources are often designed for particular tasks (e.g., part-of-speech
   tagging, named entity recognition).
2. There are plethora of different mark-up languages, which are often not fully
   compatible between systems, much less between domains.
3. Each linguistic resource may use different conceptual models. For example, there are dozens of different part-of-speech tagsets (Petrov et al., 2011;
   Chiarcos, this vol.).
4. Existing linguistic resources often do not provide precise or machine readable
   definitions of the terminology they use, thus making it difficult to reuse them
   without manual investigation.
5. It is often difficult to obtain the full metadata around the creation of a resource.
   For example, metadata about the parameters set or the number of annotators
   used may be only documented within a paper or may be given at high level but
   not on per result basis.

Constraints 1 and 3 stem from choices made early on in the design of the resource. Constraint 2 is related to the design of the resource, but can often be overcome by mapping one annotation format to another. Constraints 4 and 5 pertain to
the metadata associated with the resource.

1.1 Reuse Across Domains

Constraints 1 to 3 mostly bar combining resources and applying them to tasks that
they have originally not been created for. Naturally, there is a trade-off between
use and reuse, in particular in very domain-specific resources; the more specific
a resource, the less reusable to others. But this does not mean that the linguistic
community should not strive to facilitate reuse. The simplest scenario for which
one wants to foster reuse is the case where one wants to combine two corpora

[1] http://www.w3.org/wiki/SweoIG/TaskForces/CommunityProjects/Link
ingOpenData Retrieved: 30 November 2011

[2] http://lod2.eu/ Retrieved: 30 November 2011

[3] http://nlp2rdf.org Retrieved: 30 November 2011

that have been annotated for named entity recognition to create a larger training set for a statistical named entity recognition tool. Currently, the different conceptual models and mark-ups of corpora are not compatible with each other making this a difficult task, if possible at all.[4] In another scenario, one could use a resource that was for example developed for named entity recognition to train a part of speech tagger (as such resources often also contain part-of-speech information). As there is a particular annotation format used in the resource, which will rarely be the same as the annotation format that a tool supports, some form of data conversion is needed.

1.2 Reuse Across Communities

Constraints 4 and 5 bar reuse of resources by external parties as they make it more difficult for them to assess the data model, provenance of the data and quality of the data. This is a particularly discouraging issue for researchers or users who are not from the NLP community, but who would like to reuse linguistic resources for their applications. Because they are not familiar with the corpora, the threshold to start using them is high. This is mitigated by the fact that it is difficult for outsiders to assess the quality of the data, in particular with data that is generated (semi-)automatically.

Adopting a Linked Data approach can overcome these constraints. In addition to this, a Linked Data approach for linguistic resources may enrich existing annotations and create new opportunities for NLP tools that benefit from background knowledge.

In the remainder of this paper, an overview of how Linked Data approaches can be applied to linguistic resource reuse is given. This is done by going through the tasks necessary to convert two named entity recognition resources to Linked Data. But first, a brief overview of Linked Data is given in the next section.

2 Linked Data

The term 'Linked Data' is used to both describe the set of best practices for publishing and connecting structured data on the Web and to refer to the collection of data sets that have been published in this way so far. It gained traction with the Linked Open Data community project which promotes the publication of open data sets as RDF on the Web and by specifying links between instances of the different data sources. At the core of Linked Data lie the following four rules (Bizer et al., 2009a):

1. Use URIs as names for things
2. Use HTTP URIs so that people can look up those names.

---

[4] I mean possible without re-annotating the entire corpus.

3. When someone looks up a URI, provide useful information, using the standards (RDF*, SPARQL)
4. Include links to other URIs. so that they can discover more things.

The number of data sets that is published has grown explosively over the past four years, and with it, use cases and applications have started to become available.

3 Tasks

In this section, the different tasks necessary to convert linguistic resource data to Linked Data are detailed. As a running example, the CoNLL 2003 language-independent named entity shared task (Tjong Kim Sang and Meulder, 2003) and the ACE 2005 entity detection and recognition task (Consortium, 2005) are used. The task of named entity recognition is taken as an example because it is a fairly well-understood and successful task in NLP. But even for this task, there are many differences in various resources. In the CoNLL shared task, the following four named entity types are discerned: PERSON, LOCATION,ORGANISATION, MISCELLANEOUS.

In the ACE 2005 annotation guidelines, the following seven entity types are discerned: FACILITY, GEO-POLITICAL ENTITY, LOCATION, ORGANISATION, PERSON, VEHICLE and WEAPON. In ACE, each of the seven main entity types also has several subtypes, for example *Facility* has subtypes *Airport*, *Building-Grounds*, *Path*, *Plant*, and *Subarea-Facility*.

Furthermore, in the CoNLL data set, the data representation is quite simple to understand and work with; each token is represented on a single line, followed by its part-of-speech tag and whether it is a named entity or not in IOB formatting. The ACE data is represented in XML, and contains, besides part-of-speech tag and entity information, also information about the type of reference the entity makes to something in the world and whether there are co-referring mentions. For reasons of space, only the main entity types and their mark-up are discussed.

3.1 Representation of Linguistic Annotations in RDF

Representing linguistic annotations in RDF is mostly a straight-forward transformation, and it is the least difficult of the conversion process for reuse. This step is successfully carried out if all information contained in the original format is represented in RDF. Linked Data is represented as an Subject-Predicate-Object structure, where the objects and subjects are either resources (e.g., objects that can be grounded in some ontology) or literals (e.g., simple strings or integers). It is very well possible to represent an annotated linguistic resource such as the CoNLL data set in RDF. First, one needs to create an ontology that specifies for each concept what it denotes. For the CoNLL example, our ontology would contain concepts such as words and sentences, the part of speech tagset, the chunk tagset and the named entity types. Then every sentence in the corpus can be represented as an RDF triple that contains a sentence-ID and the sentence text, for example we can create a triple `<conll:sent1>` `<rdfs:label>` `'U.N. official Ekeus heads for Baghdad.'`. We then represent each word in the sentence as a set of RDF triples that contain the word, its position and links it to the sentence triple. We can then add triples for the part-of-speech tags that link each tag to each word, translate the chunk tags into RDF triples that tell us for each sentence at which position in the sentence a chunk starts and where it ends and likewise for each named entity where every entity mention starts and ends.

For the ACE 2005 data set, roughly the same process can be followed, although there the annotation already more extensive, it for example already encodes sentence IDs and the position of a word by character offset in the sentence. This can all be encoded in RDF triples.

When an RDF representation is created for every major element of the lexical resource, this first step in making linguistic resources more reusable has been achieved. The main requirement here is that a data format is used that is standardised and RDF is but one option. However, using RDF enables one to utilise semantic web technologies, which aid in dealing with the following task.

3.2 Mapping Annotations

The fact that two or more resources share the same format, does not automatically mean that they are integrated. Most resources will not be annotated following the exact same conceptual model, hence it is necessary to create mappings between the conceptual models of the resources as to have a unified annotation model. This task is carried out successfully when for each of the concepts in the one resource, it is clear what its counterpart is in the other resource. Much can be learnt from the ontology mapping work done in the Semantic Web (e.g., Euzenat et al. 2011).

As mentioned at the beginning of this section, there is a discrepancy between the number of entity types that is discerned in the ACE 2005 and CoNLL tasks. Part of this can be explained that the CoNLL shared task explicitly addresses 'named entities', and the ACE task only specifies 'entities', but the line between them is blurry. Indeed an entity type weapon may not always denote a named entity, but 'AK-47' can possibly be recognised as a name. One of the ACE facility subtypes is 'airports', in the CoNLL annotation, airports such as "Zaventem" are treated as locations. Another interesting issue is that of geo-political entity in ACE05, as this denotes terms such as "London" and "France" which can be locations, but they can also be used to refer to particular organisations (e.g., *France will combat market speculation*) or groups of persons (e.g., *London cleans up from riots*). In the CoNLL annotation, these would all be considered locations. In both the linguistic and Linked Data community, a satisfying high-level representation for such cases has not been found yet (cf. Recasens et al. 2011 and Halpin et al. 2010 respectively).

On a more fine-grained level, one also finds differences in entity annotations between corpora. In some corpora one will find for example that salutations, such as *Mr* and *Mrs*, are annotated as part of a person entity, in ACE and CoNLL they do not belong to the entity.

In general, it is easier to map more specific annotations such as those of ACE, to more general annotations, such as those of CoNLL, although this does mean that one cannot make use of the specificity of the ACE annotations anymore. In order to map the other way round, more information is needed, which is oftentimes not available in the more coarse-grained resource. There is no solution to this problem as yet. Possible solutions to this may be found in bootstrapping from the more specific resource to try to obtain finer granularity in the annotations in the lesser specific resources.

In order for interoperability and for a like-minded representation of named entities as Linked Data, it is important that such differences are either reconciled or mapped.

3.3 Grounding Annotations in Linked Data

When mapping annotation schemes and representing information as RDF, one has successfully made the transition to more reusability. However, Linked Data promises a large interlinked cloud of information. To ensure that linguistic resources find their way to other users and to enable tight integration with other resources, it is necessary to not only create mappings among linguistic resources but to also create mappings to other resources. For named entities one can very well imagine that mappings between entity types and DBpedia (Bizer et al., 2009b) are created and for part-of-speech tags that mappings are created that link to WordNet (Fellbaum, 1998).

For named entity recognition, one can also imagine grounding the entities in the LOD cloud[5], for example by mapping locations to GeoNames,[6] persons to EntityPedia[7], organisations to WikiCompany[8] or general entities to DBpedia.[9] Numerous implementations for the task of entity linking are available (e.g., Hellmann et al., this vol.), but in general, this is not a trivial task and the tradeoff between costs and linking quality should be carefully investigated.

---

[5] http://richard.cyganiak.de/2007/10/lod/ Retrieved: 30 November 2011

[6] http://www.geonames.org Retrieved: 30 November 2011

[7] http://entitypedia.org/ Retrieved: 30 November 2011

[8] http://wikicompany.org/ Retrieved: 30 November 2011

[9] http://dbpedia.org/ Retrieved: 30 November 2011

3.4 Definition of Linguistic Resource Metadata

Besides converting the content of linguistic resources to RDF, it is also important to convert the meta-data about the resource to a machine-readable format. At the minimum level, this meta-data describes the conceptual model of the resources, i.e., describing the different elements of the resource. Ideally, this also includes information about how the data was collected, when it was annotated, and mappings to previous versions. As more (semi-)automatically annotated resources are being shared, quality assessment becomes more important. Before one decides to reuse a particular resource, it is good to know what the quality of this resource exactly is, so one can take this into account when working with it, preferably on instance level (e.g., one can add an RDF triple for each automatically generated annotation that indicates the confidence the annotation system has in the classification).

4 Related Work

Many initiatives have been undertaken to facilitate reuse of linguistic resources. Most of these have focused on creating standards for annotations (e.g., Pustejovsky et al., 2010) or making different annotation schemes interoperable, such as the mapping of different part-of-speech tagsets (Teufel, 1997; Petrov et al., 2011), and the development of RDF-based interchange formats, which was already mentioned in Section 1. Such an interchange format should facilitate exchange of annotations between different NLP tools. As currently the draft of the first version of the NLP Interchange Format is produced and the project is still underway, its impact on the NLP community and its resources is yet to be awaited.

A nice example of the added benefit of using semantic web technology for Natural Language Processing tasks is Mika et al. (2008), who create a mapping between the CoNLL NER tagset and Wikipedia, and subsequently use this to improve the NER process.

5 Conclusion

In this contribution, the tasks and goals necessary to make linguistic resources reusable using semantic web technologies have been outlined. The main issues preventing reuse are standardisation and reconciliation of conceptual models. Solutions to these issues are sought in 1) converting existing annotation formats to RDF, 2) mapping annotations to each other and/or to a universal annotation scheme, 3) grounding the resources in the linked open data cloud, and 4) defining the metadata. These steps are the necessary prerequisites to facilitate reuse, but for reuse to be achieved it is imperative that the linguistics and semantic web community collaborate and share their expertise.

64 Marieke van Erp

References

Bizer C, Heath T, Berners-Lee T (2009a) Linked data - the story so far. International Journal on Semantic Web and Information Systems (IJSWIS) 5(3):1–22

Bizer C, Lehman J, Kobilarov G, Auer S, Becker C, Cyganiak R, Hellmann S (2009b) DBpedia - A crystallization point for the web of data. Web Semantics: Science, Services and Agents on the World Wide Web 7(3):154–165

Chiarcos C (this vol.) Interoperability of corpora and annotations. pp 161–179

Chiarcos C, Hellmann S, Nordhoff S (this vol.) The Open Linguistics Working Group of the Open Knowledge Foundation. pp 153–160

Consortium LD (2005) ACE (Automatic Content Extraction) English Annotation Guidelines for Entities version 5.6.1

Eckart K, Riester A, Schweitzer K (this vol.) A discourse information radio news database for linguistic analysis. pp 65–75

Euzenat J, Meilicke C, Stuckenschmidt H, Shvaiko P, Trojahn C (2011) Ontology alignment evaluation initiative: Six years of experience. Journal on Data Semantics 15:158–192

Fellbaum C (ed) (1998) WordNet: An Electronic Lexical Database. The MIT Press

Halpin H, Hayes PJ, McCusker JP, McGuinness DL, Thompson HS (2010) When owl:sameAs isn't the same: An analysis of identity in linked data. In: The 9th International Semantic Web Conference (ISWC 2010), Shanghai, China, pp 305–320

Hellmann S, Stadler C, Lehmann J (this vol.) The German DBpedia: A sense repository for linking entities. pp 181–189

Mika P, Ciaramita M, Zaragoza H, Atserias J (2008) Learning to tag and tagging to learn: A case study on Wikipedia. IEEE Intelligent Systems 23(5):26–33

Petrov S, Das D, McDonald R (2011) A universal part-of-speech tagset. arXiv:1104.2086v1

Pustejovsky J, Lee K, Bunt H, Romary L (2010) ISO-TimeML: An international standard for semantic annotation. In: Proceedings of LREC 2010, pp 394–397

Recasens M, Hovy E, Martí MA (2011) Identify, non-identity, and near-identity: Addressing the complexity of coreference. Lingua pp 1138–1152

Teufel S (1997) A support tool for tagset mapping. arXiv:cmp-lg/9506005v2

Tjong Kim Sang EF, Meulder FD (2003) Introduction to the conll-2003 shared task: Language-independent named entity recognition. In: Proceedings of CoNLL-2003, Edmonton, Canada, pp 142–147

Windhouwer M, Wright SE (this vol.) Linking to linguistic data categories in ISO-cat. pp 99–107

A Discourse Information Radio News Database for Linguistic Analysis

Kerstin Eckart, Arndt Riester, and Katrin Schweitzer

Abstract In this paper we present DIRNDL, an annotated corpus resource comprising syntactic annotations as well as information status labels and prosodic information. We introduce each annotation layer and then focus on the linking of the data in a standoff approach. The corpus is based on data from radio news broadcasts, i.e. two sets of primary data: spoken radio news files and a written text version which sometimes deviates from the actual spoken data. We utilize a generic relational database management system to bridge the gap between the deviating primary data as well as between the different properties of the annotation levels. We show how the resource can support data extraction concerning the interface between information status, syntax and prosody.

1 Introduction

We present the DIRNDL corpus (Discourse Information Radio News Database for Linguistic analysis), an annotated resource of news broadcasts from Deutschlandfunk, a German radio station, prepared for the investigation of the interfaces between prosody, information status and syntax.[1] The database contains audio files (approx. 5 hours of speech; 9 speakers: 5m, 4f), which were annotated for pitch accents and prosodic boundaries following GToBI(S) (Mayer, 1995). Furthermore, it comprises a treebank based on the written manuscripts of the news (3221 sentences), which were annotated for referential information status (given-new distinction), according to Riester et al. (2010). The two types of data are aligned in a generic relational database management system described in Eckart et al. (2010).

Kerstin Eckart · Arndt Riester · Katrin Schweitzer
Universität Stuttgart, Institut für Maschinelle Sprachverarbeitung, Azenbergstr. 12, 70174 Stuttgart
e-mail: {eckartkn,arndt.riester,katrin-schweitzer}@ims.uni-stuttgart.de

[1] News broadcasts from 25-27/03/2007; downloaded from http://www.dradio.de.

C. Chiarcos et al. (eds.), *Linked Data in Linguistics*,
DOI 10.1007/978-3-642-28249-2_7, © Springer-Verlag Berlin Heidelberg 2012

2 Two Annotation Pipelines

There exist two primary data sets: spoken data and a slightly deviating written version. The annotation layers are the results of two different processing pipelines: one from the written primary data to recursive information status labels, and the other from the spoken primary data to prosodic annotations.

2.1 Workflow Towards Information Status Annotations

The written manuscripts of the news were parsed with the XLE system and the German LFG-grammar by Rohrer and Forst (2006). The resulting constituent trees[2] were converted into TIGER-XML using TIGERRegistry (Lezius et al., 2002). A sample is shown in Fig. 1.

Information status was annotated to syntactic nodes. We used the SALTO/SALSA tool (Burchardt et al., 2006) which allows for a free definition of annotation labels (in our case, information status labels), and which takes TIGER-XML as input, see Fig. 2. Information status (Prince, 1981, 1992) describes the degree of givenness of (referential) expressions. On a slightly different interpretation, it classifies terms as to whether they are anaphoric, inferable, deictic, or discourse-new. Notions closely related to information status are *salience*, *accessibility* and *cognitive status*.

Information status forms a subfield of information structure theory, since it is usually confined to referential expressions and furthermore leaves aside aspects of contrastive focus. For the annotation of the DIRNDL corpus, we made use of the scheme defined in Riester et al. (2010), which is particularly suited to handle multiple embeddings, which are very frequent in news text, see Fig. 2. The scheme has been shown to reach an inter-annotator agreement of $\kappa = .66$ for the full scheme of 21 categories and $\kappa = .78$ for a core scheme of 6 main categories.

2.2 Workflow Towards Prosodic Annotations

The spoken primary data set was automatically segmented into words, syllables and phonemes using forced alignment (Rapp, 1995). Pitch accents and prosodic boundaries were manually labelled according to GToBI(S) (Mayer, 1995). Word level annotations were mapped to the syllable-based prosodic labels using Festival (Taylor et al., 1998).

Fig. 3 shows the representation of time-aligned word boundaries, combined with phrase boundaries and pitch accents, all included as annotations in the corpus. While some words can be unaccented (e.g. the determiners and prepositions in Fig. 3),

---

[2] We always used the parses with the highest rank.

```
<s id="s7">
 <graph root="s7_500">
  <terminals>
    ...
   <t id="s7_6" word="die" pos="D[std]"/>
   <t id="s7_7" word="Tuer" pos="N[comm]"/>
   <t id="s7_8" word="zu" pos="P[pre]"/>
   <t id="s7_9" word="Verhandlungen" pos="N[comm]"/>
   <t id="s7_10" word="mit" pos="P[pre]"/>
   <t id="s7_11" word="Teheran" pos="NAME"/>
    ...
  </terminals>
  <nonterminals>
    ...
   <nt id="s7_511" cat="DPx[std]">
    <edge label="--" idref="s7_6"/>
    <edge label="--" idref="s7_512"/>
   </nt>
   <nt id="s7_517" cat="NP">
    <edge label="--" idref="s7_9"/>
   </nt>
    ...
  </nonterminals>
 </graph>
  ...
</s>
```

Fig. 1 Sample phrase in TIGER-XML format

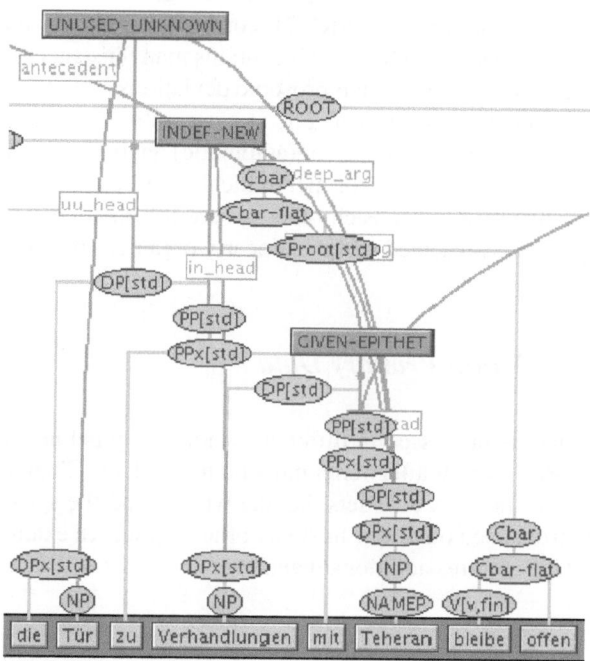

Fig. 2 Annotation of the phrase from Fig. 1 in SALTO/SALSA

others, especially compounds, may carry more than one pitch accent. Such cases
are represented as complex accents in the resource.

Fig. 3 Sample phrase with prosodic annotations (time stamps denoting word boundaries, words, phrase boundaries, pitch accents)	54.480000 die	NONE	NONE
	54.790000 T"ur	NONE	|H∗|
	55.060000 zu	NONE	NONE
	55.750000 Verhandlungen	NONE	|!H∗|
	55.890000 mit	NONE	NONE
	56.430000 Teheran	%	|!H∗L|
	57.180000 bleibe	NONE	|L∗|
	57.540000 offen	%	|H∗L|

2.3 Differences in Annotation Structure

There are two major differences between spoken and written language which have
an influence on annotation decisions. First, speech has a temporal dimension. Every
word token and every tonal event occurs at a specific time point or interval. Written
language obviously lacks this temporal determination since it can be read at varying
speed. A related issue, which, for lack of space, we cannot discuss in detail, is the
fact that written language is often underspecified as regards its pragmatic impact. We
want to mention, however, that the DIRNDL corpus is a good resource for studying
meaning specification via prosody, since it contains many instances of repetitions of
identical news features showing small prosodic deviations.

Second, as we pointed out in Sect. 2.1, to systematically annotate information
status within complex news language, an (automatic) analysis of syntactic structure
is indispensable in order to highlight hierarchical relations. As referential expressions are often embedded inside each other, so are information status labels. This
cannot be adequately represented within a linearly organised phonetic analysis tool.

2.4 Deviations Within Primary Data

When primary data is processed in different annotation pipelines, conflicting tokenizations may arise, which afterwards must be merged, cf. Chiarcos et al. (2009).
In our case, the two primary data sets, i.e. the written and the spoken one, already
slightly deviate from each other due to slips of the tongue, see example (1), or other
modifications. This requires additional handling.

(1)   **Bundeskanzler** Köhler hat das **ich korrigiere** Bundespräsident Köhler hat
       das Gesetz zur Gesundheitsreform unterschrieben
       '(Chancellor Köhler, correction) Federal President Köhler signed the bill on
       the health care reform'

As stated above, the processing of the data in different pipelines introduces even
more deviations. Tokens in the prosodic pipeline refer to actually pronounced items.
This leads to an inhomogeneous treatment of punctuation symbols. Hyphens, like
in *EU-Außenbeauftragter* ('EU High Representative') are not pronounced and disappear in the transcriptions of the speech data, while the comma symbol in a nu-
meric expression like *6,9* becomes a token of its own and is transcribed as the word
Komma.

Choosing only one of the primary data sets means information loss in processing.
On the one hand, slips of the tongue create problems for the parser. On the other
hand, they have an influence on prosody. It is therefore not advisable to cut out parts
of the speech data. To handle the differences between the primary data sets and
the differences between the outputs of the processing pipelines we introduce links
between the tokens created by each pipeline. That way, we are able to keep as much
information as possible in the corpus and are even able to extract data for the study
of specific phenomena such as the prosody of slips of the tongue.

3 A Generic Database Management System

Our database[3] is able to handle different data sets like primary data, metadata and
linguistic annotations, cf. Eckart et al. (2010). It meets the requirements for a resource like DIRNDL, as it is *extensible*, *theory-independent* and supports the versioning of annotations within a processing pipeline. Extensibility is important, as
it allows to include more data sets into our resource at a later point. This is easily
achieved since the generic data structures of the database allow the inclusion of new
kinds of data without changes to the schema.

The database is conceptually divided into two different layers. At the *macro-
scopic* level each data set is represented as an object. Metadata about these objects
are provided by sorting each object into a group (e.g. *corpus* for a set of primary
data, or *analysis* for the result string produced by an analysis tool) and assigning it a
type (e.g. *speech* for an object of group *corpus*). Versioning information is included
in the form of a creation date. Other optional attribute-value-pairs can be used to
add metadata, like author information etc.

Objects which contain further internal structure, such as a parse tree represented
as a bracketed string, can be represented as graphs at the *microscopic* level. The
data structures on the microscopic layer are mainly typed nodes and edges. The

---

[3] Implemented as PostgreSQL relational database system. http://www.postgresql.org

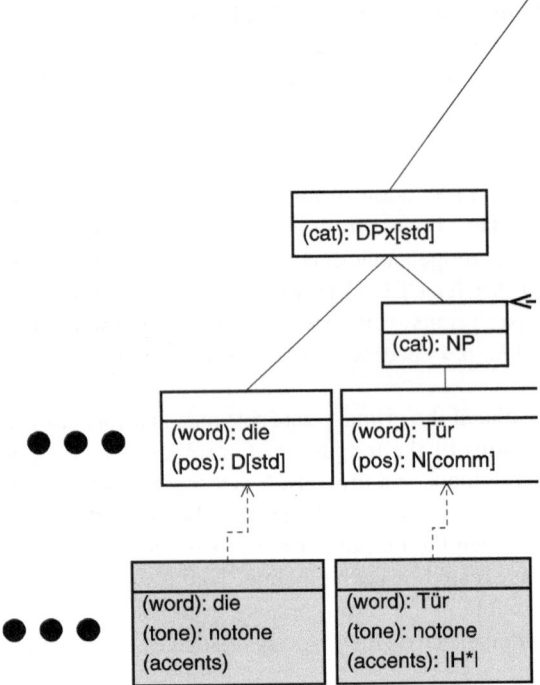

Fig. 4 Linked annotation graphs in the database for the sentence: *Der EU-Außenbeauftragte Solana betonte, die Tür zu Verhandlungen mit Teheran bleibe offen.* ('The EU High Representative Solana stressed that the door for negotiations with Teheran remained open.')

schema is enhanced with structures based on the Graph Annotation Format (GrAF),[4] providing feature structures to annotate nodes and edges. For DIRNDL, we make use of GrAF-based data structures for all annotation layers.

4 Linking Annotations

With respect to each of the pipelines, the GrAF-based data-structures provide a standoff approach to the representation of each annotation layer in the database. The prosodic annotations are based on the spoken and time-aligned primary data set, the syntactic analyses from XLE are based on the written primary data set and the labels refer to the nodes in the constituent trees. Each layer is interpreted as a graph of a different type in the database:

---

[4] GrAF ist the XML serialization of the upcoming ISO-Standard LAF (Linguistic Annotation Framework, ISO/DIS 24612). LAF proposes a theory-independent exchange format based on a standoff approach.

- Each constituent tree (a sentence) is a graph; see the white nodes and the edges marked by continuous lines in Fig. 4.
- An information status graph contains all that refer to the same constituent tree; see the dark grey nodes in Fig. 4.
- A prosody graph comprises a complete broadcast rather than a single sentence; compare the light grey nodes in Fig. 4.

While the syntactic graphs include nodes and edges, prosody and information status are represented by unconnected graphs. They only consist of nodes. The prosody nodes become sequential when annotated with time-stamps while the information status graphs represent hierarchical information.

To prevent information loss, all information available in the results of the annotation pipelines is kept in the database in the form of annotations. This does not only comprise linguistic information like part-of-speech tags, but also the administrative information of the original SALSA-XML files (e.g. identifiers). The relations connecting the information status labels with their respective constituent trees are explicitly included in the SALSA output file. They are represented in the database as link edges between their respective information status and syntactic graphs; see the dashed edges in Fig. 4.

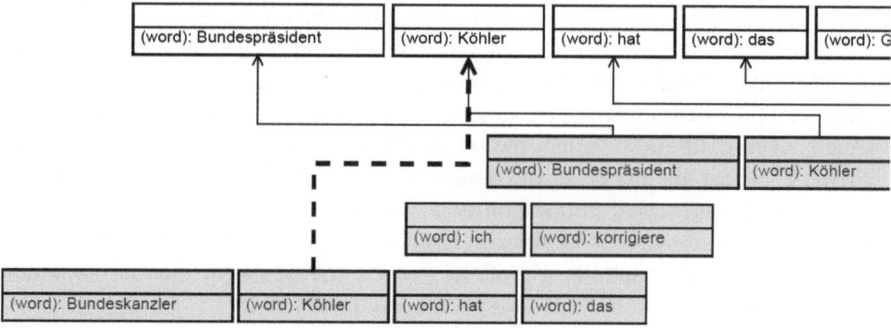

Fig. 5 Links for example (1); tokens from written version (white), tokens from spoken version (grey), primary links, secondary link (dashed).

As a last step, we integrate the annotations of the two pipelines, by utilizing a semi-automatic mapping at token level, i.e. between the terminal nodes of the syntactic and the prosodic graphs. The algorithm takes a file with the terminal nodes from each data set as input and reads the first node from both files; if the tokens are identical or can be systematically mapped, like in the case of punctuation symbols ($|6,9|$ vs. $|6|Komma|9|$; $|EU\text{-}Au\betaenbeauftragter|$ vs. $|EU|Au\betaenbeauftragter|$), a link between the nodes can be inserted into the database. If the algorithm fails to map the tokens[5] the algorithm stops and prints out the tokens to the user. Then the user excludes problematic tokens from the input files and starts the mapping script

---

[5] The procedure is rather restrictive here to avoid mapping mismatches.

again. The user may now decide where to manually insert additional links. This is often the case with slips of the tongue, like in example (1). By also assigning types to the link edges, different mismatches can be identified and explicitly included in or excluded from queries, see Fig. 5.

5 Querying Information Status, Syntax and Prosody

As annotations from all layers are related via links, any combination of annotations can be used in a query. This means, however, that queries may become relatively complex, because all layers that must be included into or excluded from the query result need to be explicitly specified. In the trade-off between genericity and ease of query formulation, we have opted for the former.

In the following, we briefly describe a simple query, which is meant to demonstrate the interplay of the three linguistic levels of prosody, discourse (information status) and syntax. We want to investigate the prosodic realization of phrases consisting of exactly two words (in the written tokenization) which carry an information status label. This is formulated in the form of an SQL query, an excerpt of which is shown in Fig. 6. We run the query on a one-day subset of the data which at the time of publication of this paper has been integrated into the database.

For this query we generate the database table is_syn_p, which contains all information status labels, their corresponding text phrases and the respective accent patterns found when following the links from the tokens of the written to the tokens of the spoken dataset. We select the phrases which comprise two words (e.g. *mit Teheran*), see last line in Fig. 6, and obtain the results in Fig. 7, which show that the percentage of unaccented phrases on two-word expressions decreases along with the degree of salience: 14% of the coreference anaphora (GIVEN) are unaccented, 7% of the bridging anaphora, 4% of the generic terms, 2% of the discourse-new definites (UNUSED) and none of the specific indefinites (NEW).[6]

6 Availability

As the data structures of our resource are based on GrAF, which is already an exchange format, we intend to export the annotation layers in the GrAF XML format to make them available for research purposes. Figures 8 and 9 show parts of a GrAF-export for the sentence shown in Fig. 4: Fig. 8 is information on the UNUSED-KNOWN node from the information status graph, and Fig. 9 shows the representation

---

[6] The information status categories have been simplified in the following way: GIVEN subsumes GIVEN-EPITHET, -REPEATED, -SHORT; BRIDGING includes BRIDGING and BRIDGING-TEXT; UNUSED stands for UNUSED-KNOWN and UNUSED-UNKNOWN; NEW subsumes INDEF-NEW and INDEF-PARTITIVE; GENERIC combines INDEF-GENERIC and UNUSED-TYPE. For details, see Riester et al. (2010) and Baumann and Riester (to appear).

```
SELECT
    is_syn_p . syn_s_num ,
    is_syn_p . is_label ,
    is_syn_p . phrase ,
    is_syn_p . accent_sequence
FROM
    is_syn_p ,
    sentences
WHERE
    is_syn_p . syn_s_num = sentences . s_num
    AND
    is_syn_p . syn_phrase_length = 2;
```

Fig. 6 An excerpt of the SQL query discussed in Sect. 5

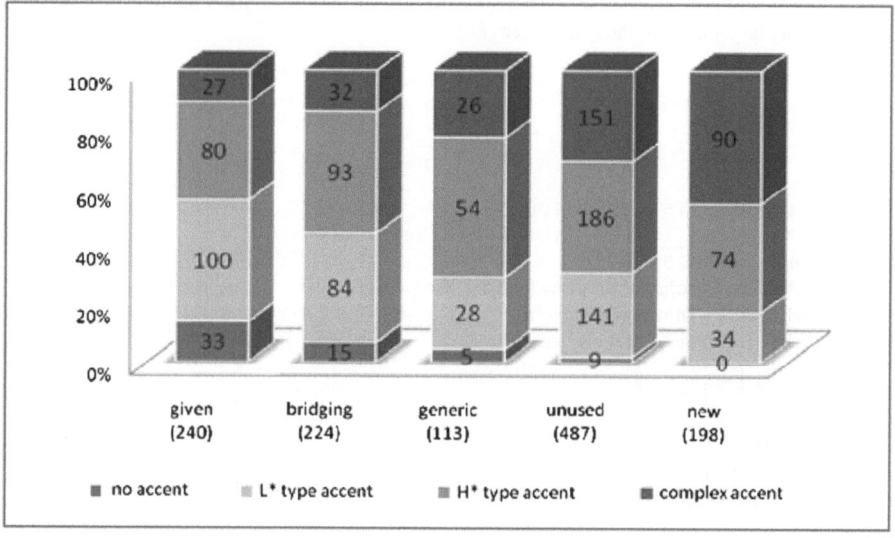

Fig. 7 The pitch accents on two-word terms depending on information status. Results of the query in Fig. 6.

of the respective target node in the syntax tree (a *DP* node) and one of the terminal nodes in the phrase (*Tür*). The generic GrAF XML format is not only intended to be convertible into different tool input-formats but also into other graph-based generic formats, such as PAULA XML (Dipper, 2005). At the moment, different researchers also address the development of RDF and OWL linearizations of such graph-based generic formats: Cassidy (2010), for example, proposed an RDF linearization of GrAF, and Chiarcos (this vol.) developed an OWL/RDF linearization of PAULA XML. Through these recent developments, our approach is linked to the creation

```
<node xml:id="n215324_24941" />
<a ref="n215324_24941" label="a1_is_scheme">
 <fs xml:id="fs367562">
  <f value="UNUSED-UNKNOWN" name="name" />
 </fs>
</a>
<edge to="n151089_19406" from="n215324_24941"
      xml:id="e162443" />
<node xml:id="n151089_19406" />
<a ref="n151089_19406" label="xle_nonterminal">
 <fs xml:id="fs240027">
  <f value="DP[std]" name="cat" />
 </fs>
</a>
```

Fig. 8 Samples of DIRNDL in GrAF format: UNUSED-KNOWN node from the information status graph, and its target node in the syntax tree (*DP*).

```
<node xml:id="n151049_19406" />
<a ref="n151049_19406" label="xle_terminal">
 <fs xml:id="fs239987">
  <f value="N[comm]" name="pos" />
  <f value="7" name="seq" />
  <f value="Tuer" name="word" />
 </fs>
</a>
<edge to="n151076_19406" from="n151089_19406"
      xml:id="e92409" />
<edge to="n151067_19406" from="n151089_19406"
      xml:id="e92410" />
```

Fig. 9 Samples of DIRNDL in GrAF format: A terminal node (*Tür*) of the *DP* node referred to in Fig. 8.

of interoperable representations of multi-layer corpora by means of Semantic Web technologies, and to provide corpora as Linked Data.

Acknowledgements This work was funded by the German Research Foundation DFG, via the Collaborative Research Centre SFB 732 *Incremental Specification in Context*.

References

Baumann S, Riester A (to appear) Referential and Lexical Givenness: Semantic, Prosodic and Cognitive Aspects. In: Elordieta G, Prieto P (eds) Prosody and Meaning Interface Explorations, De Gruyter Mouton, Berlin

Burchardt A, Erk K, Frank A, Kowalski A, Padó S (2006) SALTO: A Versatile Multi-Level Annotation Tool. In: Proceedings of the Fifth International Conference on Language Resources and Evaluation (LREC), Genoa, Italy

Cassidy S (2010) An RDF realisation of LAF in the DADA annotation server. In: Proceedings of the 5th Joint ISO-ACL/SIGSEM Workshop on Interoperable Semantic Annotation (ISA-5), Hong Kong

Chiarcos C (this vol.) Interoperability of corpora and annotations. pp 161–179

Chiarcos C, Ritz J, Stede M (2009) By all these lovely tokens... Merging Conflicting Tokenizations. In: Proceedings of the Third Linguistic Annotation Workshop, Association for Computational Linguistics, Suntec, Singapore, pp 35–43

Dipper S (2005) XML-based Stand-off Representation and Exploitation of Multi-Level Linguistic Annotation. In: Proceedings of Berliner XML Tage 2005 (BXML 2005), Berlin, pp 39–50

Eckart K, Eberle K, Heid U (2010) An Infrastructure for More Reliable Corpus Analysis. In: Proceedings of the Workshop on Web Services and Processing Pipelines in HLT: Tool Evaluation, LR Production and Validation (LREC'10), Valletta, Malta, pp 8–14

Lezius W, Biesinger H, Gerstenberger C (2002) TIGERRegistry Manual. Tech. rep., IMS Stuttgart

Mayer J (1995) Transcription of German Intonation. The Stuttgart System. URL http://www.ims.uni-stuttgart.de/phonetik/joerg/labman/STGTsystem.html, ms

Prince EF (1981) Toward a Taxonomy of Given-New Information. In: Cole P (ed) Radical Pragmatics, Academic Press, New York, pp 233–255

Prince EF (1992) The ZPG Letter: Subjects, Definiteness and Information Status. In: Mann W, Thompson S (eds) Discourse Description: Diverse Linguistic Analyses of a Fund-Raising Text, Benjamins, Amsterdam, pp 295–325

Rapp S (1995) Automatic Phonemic Transcription and Linguistic Annotation from Known Text with Hidden Markov Models – An Aligner for German. In: Proceedings of ELSNET Goes East and IMACS Workshop "Integration of Language and Speech in Academia and Industry" (Russia)

Riester A, Lorenz D, Seemann N (2010) A Recursive Annotation Scheme for Referential Information Status. In: Proceedings of the Seventh International Conference on Language Resources and Evaluation (LREC), Valletta, Malta, pp 717–722

Rohrer C, Forst M (2006) Improving Coverage and Parsing Quality of a Large-scale LFG for German. In: Proceedings of the Fifth International Conference on Language Resources and Evaluation (LREC), Genoa, Italy

Taylor P, Black AW, Caley R (1998) The Architecture Of The Festival Speech Synthesis System. In: Proceedings of the Third ESCA Workshop in Speech Synthesis, pp 147–151

Integrating Treebank Annotation and User Activity in Translation Research

Michael Carl and Henrik Høeg Müller

Abstract The Center for Innovation of Translation and Translation Technology (CRITT) environment at Copenhagen Business School (CBS) draws on primarily two types of NLP resources, namely treebanks and the logging of user activity data (UAD) during text production and translation activities, in order to do research into the cognitive processes that lie behind translation activity. In this paper we make a short presentation of the Copenhagen Dependency Treebank (CDT), and elaborate how UAD is obtained and represented in Translog-II. Finally, the paper discusses some general perspectives on how process-oriented translation research methodology could benefit from the integration of UAD with structural linguistic information in the form of linguistically annotated text data.

1 Introduction

The main focus of the CRITT (Center for Innovation of Translation and Translation Technology) research environment is on the empirical and experimental study of translation processes with an applied, technological aim. Our research designs involve data elicitation methods (keystroke logging) and behavioural measuring technologies (eyetracking), as well as parallel linguistically annotated text collections (treebanks). To record user activity data (UAD), CRITT has developed the computer programme Translog-II, which logs keystrokes, mouse activities, and gaze movements during text production. With respect to treebanks, the CRITT translation research programme has devised the Copenhagen Dependency Treebank (CDT), an NLP resource which provides information about language structure and meaning on various levels. While the CDT annotates the static structure of the parallel, trans-

Michael Carl · Henrik Høeg Müller
Copenhagen Business School, Languages & Computational Linguistics,
Frederiksberg, Denmark
e-mail: {mc.isv,hhm.isv}@cbs.dk

C. Chiarcos et al. (eds.), *Linked Data in Linguistics*,
DOI 10.1007/978-3-642-28249-2_8, © Springer-Verlag Berlin Heidelberg 2012

lated texts, Translog-II provides information on how the parallel data was actually created during the translation process.

Since around the mid 90s most texts are generated by humans using a (computer) keyboard, but still there is hardly any empirical data available that is suited to investigate how translations are generated, and to uncover and describe the processes by which humans produce translations. A central aim of CRITT is to overcome this gap. In this paper, we seek to connect the two worlds of product and process annotation. The paper is structured as follows. In Section 2, it is illustrated, very briefly, how linguistic structure is annotated in CDT, including alignment. In Section 3, focus is on how UAD is structured in Translog-II. Section 4 offers some speculations about the possible benefits derived from integrating CDT and Translog-II, while in section 5 we comment on the predictability of the translators' behaviour. Finally, section 6 sums up the central points.

2 The Copenhagen Dependency Treebank

The Copenhagen Dependency Treebank (CDT) (Trautner-Kromann, 2003) is a multilingual open NLP resource which consists of linguistically annotated parallel text collections of approx. 60.000 words each for Danish, English, German, Italian, and Spanish. The CDT is based on a unified dependency annotation which includes not only syntax, but also fine-grained analyses of morphological, discourse, and anaphoric structure. Moreover, in order to extent its applicability potential to MT, the resource has an alignment system of translational equivalences that allows us to specify relations between words or word groups in the source and target language that correspond to each other with respect to meaning or function (Buch-Kromann et al., 2009).

Figure 1 shows the primary dependency tree for the sentence "Kate is working to earn money" (top arrows), enhanced with secondary subject relations (bottom arrows). The arrows point from governor to dependent, with the relation name written at the arrow tip.

CDT also allow for morphological annotation which deals with derivation and composition. The internal structure of words is encoded as a dependency tree through an operator notation scheme (Müller and Durst-Andersen, 2012).

Figure 2 illustrates how anaphora (bottom arrows) and discourse (top arrows) are annotated in CDT (Korzen and Müller, 2011). The arrows indicating anaphoric relations run from the antecedent to the anaphora, while the discourse arrows go from the top node of the governing segment to the top node of the dependent segment, the top node being typically, but not necessarily, the verb. Figure 3 plots the alignment of a Danish-English translation together with their syntactic annotation.

This multilevel annotation distinguishes CDT from other treebank projects which tend to focus on a single linguistic level, and it has the advantage of not obliging us to limit the kind of linguistic relations that can be annotated, and not having to draw precise, and often arbitrary, boundaries between morphology, syntax, and discourse.

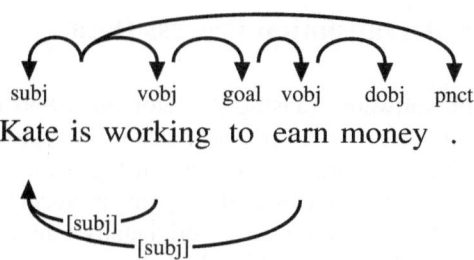

Fig. 1 Primary dependency tree (top) augmented with secondary subject dependency relations (bottom).

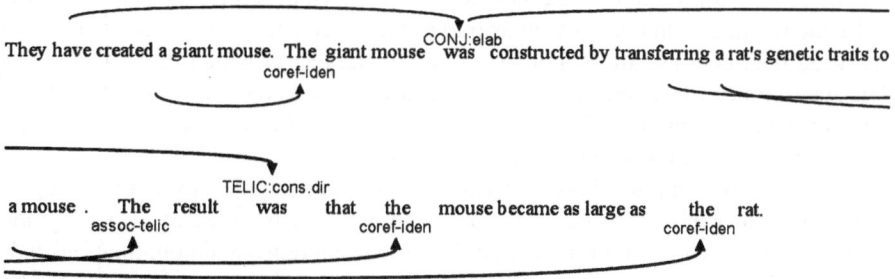

Fig. 2 Discourse and anaphoric relations in the CDT on the bottom and top lines respectively. The annotation schema is independent from the morphological, syntactic, and alignment information.

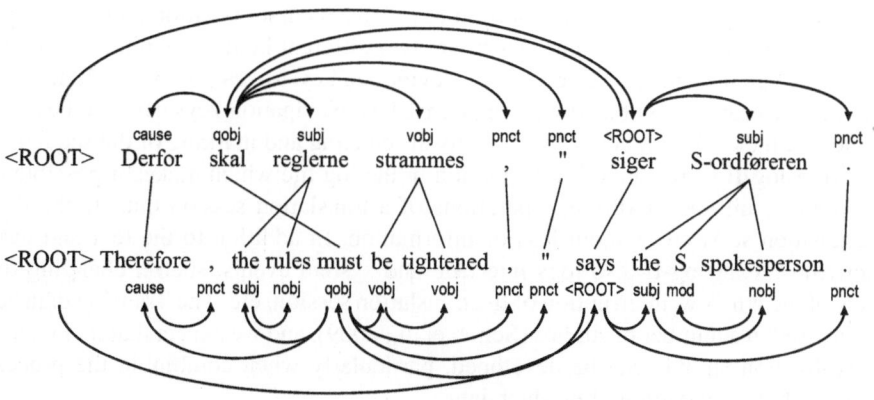

Fig. 3 The Figure shows an example of an alignment between Danish and English. For the sake of clarity only the syntactic annotation appears in the figure.

3 The Structure of Translation Process Data

The UAD acquisition software Translog-II (Jakobsen, 1999) logs keystrokes and mouse activities during text production and it also records gaze movements on the texts when connected to an eye tracker. While Translog-II can also be used for reading and writing research, we present here the logging protocol when used as a translation tool. Translog-II divides the screen horizontally into a source text window (top) and a target text window (bottom) in which the translator types her translation. The activity data is collected during a translation session for later analysis and can be replayed in a replay mode (Carl et al., 2011; Carl and Dragsted, 2011; Dragsted, 2010).

The process data also consist of three resources, the gaze sample points, fixations and keystroke information. Gaze sample points consist of screen coordinates looked at by the left and right eye, as well as pupil dilation at a particular time. In addition, the gaze data contains the windows (source or target) in which the gaze (i.e. the left eye sample) was detected. The log file also contains the location of the closest character (i.e. the cursor position of the character in the ST or TT) to the gaze sample point, and the location of that character on the screen.

Fixations are computed based on sequences of gaze samples. Fixations group together a number of near-distance eye-gaze samples which represent a time segment in which a word (or symbol) is fixated. In our current representation, fixations have a starting time, a duration, and a cursor position which refers to an index in the ST or TT.

Translog-II also logs a number of keyboard and mouse activities. We distinguish five different types of keystrokes: insertion, deletion, editing, navigation and the return key. Each of the permitted keystrokes has a slightly different representation, depending on whether or not the keystroke is visible on the screen and whether it involves a deletion or not. All insertion keys have a cursor position and screen position (X/Y, Width and Height). There are two different kinds of explicit deletions, the [Back] and the [Delete] key. Navigation keystrokes (up, down, left, right, home, end) can be combined with the Ctrl key. Navigation keys can also be combined with the Shift key, which results in selecting and marking of the sequence.

Translog-II stores the full information in the log file which makes it possible to re-produce all text modifying operations of a translation session outside the data acquisition software without loss of information. In addition to the text and gaze activities, Translog-II also logs interface and system events, such a changing the size of the window, interruption of the translation session, etc. The Translog data has been used in a number of studies (Schou et al., 2009), and we expect that many more investigation strands can be developed, particularly when combining the process data with the annotation of product data.

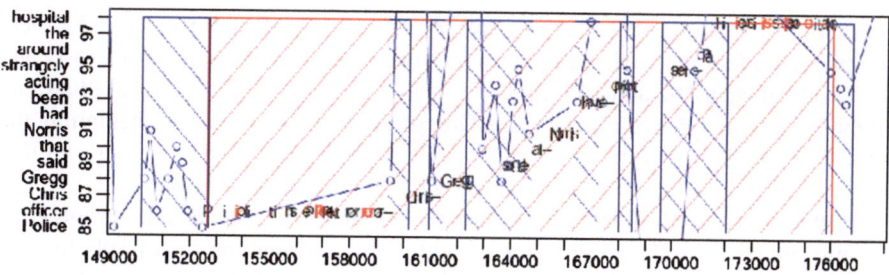

Fig. 4 Translation progression graph showing parallel reading and text production

4 Perspectives of Integrating Translog-II with CDT

In this last part of the paper, we suggest some theoretical considerations as to what could be gained from integrating Translog-II UAD with dependency annotated texts in CDT with respect to conducting basic empirical research into the cognitive processes during translation activities. In general, we expect that the combination of Tranlog-II and CDT data will facilitate enquiries into isomorphisms between, on the one hand, pauses and gaze location/duration and, on the other, linguistic layers in its broadest sense, which could reveal new insights about source text decoding, memory retrieval and encoding of textual segments, and their linguistic structure. The eye-mind hypothesis predicts a strong correlation between gaze location and mind activity, i.e. between what we are looking at and what we are thinking about. In Translation Process Research it is customary to part from the assumption that gaze location reflects the focus of attention of the translator so that longer gaze durations on the ST or TT correspond to bigger decoding (comprehension) and encoding (productions) problems, respectively, presupposing a stratificational process model of translation (Hyrskykari, 2006; Pavlovic and Jensen, 2009).

In Figure 4 we present a graph which illustrates how Translog-II records gaze data and keystrokes in a translation process (Carl and Dragsted, 2011). It presents a translation progression graph for the translation of a English source sentence into Danish:

- English source sentence:
  Police officer Chris Gregg said that Norris had been acting strangely around the hospital
- Danish translation:
  P[i]olitiins[ep]pekt[rør]ør Chris Gregg sagte at Norris havde opført sig sært på h[i]o[p[si]s]sp[o]italet

The vertical axis represents the ST and the horizontal axis the time span of 28 seconds (149-177) during which the translation took place. The red upwards hatched boxes are clusters of coherent writing activity, whereas the blue circles indicate ST fixations and the corresponding blue downwards hatched boxes symbolize fixation units. Figure 4 shows that fixations are not equally distributed over the ST. There are two large fixations units in the progression graph, i.e. between seconds 149 and 152, and between 162 and 165. In the first case, the translator's eyes moved around in the text chunk "Police officer Chris Gregg said that Norris" before beginning the new sentence, whereas in the second case the reading activity of "Gregg said that Norris had been acting strangely" occurs while typing the translation.

In the future, CRITT intends to devise computational mechanisms to integrate the process data (as shown in figure 4) with CDT information, i.e. morphological, syntactic, discourse and anaphoric, as well as alignment information as shown in figures 1 to 3.

5 Predictability of Translators' Behaviour

The distribution of gaze points and keystrokes of a single translator, of course, reflects an individual pattern of translation challenges. In order to investigate whether different translators translating the same text face similar difficulties at the same text positions, Carl and Dragsted (2011) conducted an experiment where they looked into the amount of ST gaze activity that was detected before a translation was typed, and compared the gaze durations of 5 translators. The result is plotted in Figure 5.

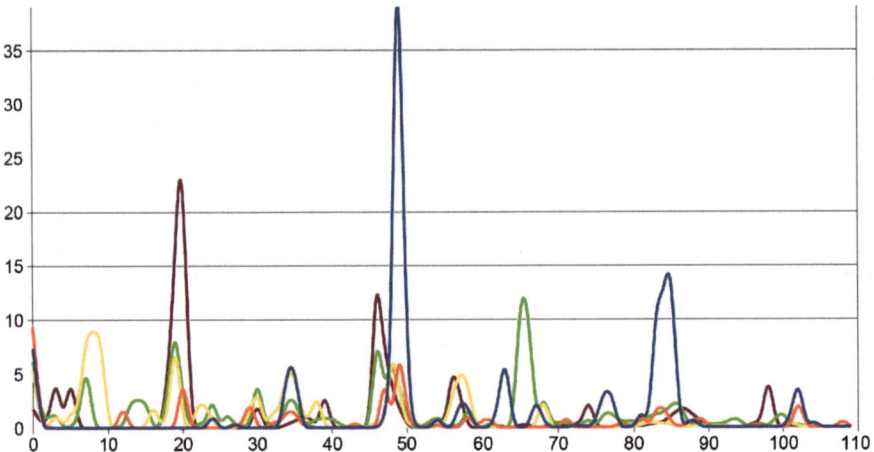

Fig. 5 Relative amount of fixation duration before/during production of the translation of ST words

The horizontal axis enumerates words in the ST. The vertical axis plots the amount of ST gaze time before typing the translation of the ST word, where each color represents a translator.

Figure 5 shows that the translators spent gaze time on close ST fragments, indicating that they have similar problems at this point in the translation.

6 Conclusion

The central problem is, however, that UAD must be directly correlated with an analysis of linguistic data in a systematic way to formulate hypotheses about the problems faced by the translators, and to establish links between sensory-motor clues and linguistic characteristics of the translated expressions. With the words of Angelone (2010) "non-articulated indicators, such as pauses and eye-fixations, give us no real clue as to how and where to allocate the uncertainty". However, if UAD is directly connected to a parallel treebank with multilevel linguistic annotation, the possibilities of systematically analyzing correlations between gaze fixations and keystrokes and underlying linguistic structure of the texts are promising to uncover the translation processes. Specifically, we assume that the integration of these two NLP resources would allow us to correlate patterns of UAD with patterns of morphological, syntactic or discourse structure. By mapping dynamic UAD on to structural treebank annotation data, behavioral factors become grounded in linguistics, and in this way we may gain a better understanding of the interconnection between text production and comprehension processes, i.e. of how cognitive activity does (or does not) correspond with the linguistic categories and complexity we are used to deal with from an analytical and theoretical perspective. The integration of product and process data opens the possibility to investigate the extent to which patterns of UAD are related to standard linguistic units in the form of phrases, sentences, etc. or whether there are correspondences between eye-gaze regression patterns and anaphoric paths, discourse structures or complex phrases.

References

Angelone E (2010) Uncertainty, uncertainty management and metacognitive problem solving in the translation task. Translation and Cognition pp 17–40

Buch-Kromann M, Korzen I, Müller HH (2009) Uncovering the lost structure of translations with parallel treebanks. Copenhagen Studies in Language 38: 199–224

Carl M, Dragsted B (2011) Inside the monitor model: Processes of default and challenged translation production. In: Contrastive Linguistics, Translation Studies, Machine Translation – What can we Learn from Each Other? workshop held in conjunction with the Conference of the German Society for Computational Linguistics and Language Technology (GSCL 2011), Hamburg, Germany

Carl M, Dragsted B, Lykke Jakobsen A (2011) On the systematicity of human translation processes. In: Proceedings of Translation Careers and Technologies: Convergence Points for the Future (Tralogy 2011), Paris, France

Dragsted B (2010) Coordination of reading and writing processes in translation. Translation and Cognition, American Translators Association Scholarly Monograph Series, Benjamins, Amsterdam/Philadelphia

Hyrskykari A (2006) Utilizing eye movements: Overcoming inaccuracy while track-
   ing the focus of attention during reading. Special issue: Attention aware systems
Jakobsen A (1999) Logging target text production with translog. In: Hansen G (ed)
   Probing the process in translation: methods and results, Copenhagen Studies in
   Language, vol 24, Samfundslitteratur, Copenhagen, pp 9–20
Korzen I, Müller HH (2011) The copenhagen dependency treebank. Forskellige
   niveauer samme relationer
Müller HH, Durst-Andersen P (eds) (2012) Ny forskning i Grammatik. Odense Uni-
   versitetsforlag, Odense
Pavlovic N, Jensen KTH (2009) Eye tracking translation directionality. In: Pym A,
   Perekrestenko A (eds) Translation Research Projects 2, Intercultural Studies
   Group, Tarragona, pp 93–109. http://isg.urv.es/publicity/isg/
   publications/trp_2_2009/index.htm
Schou L, Dragsted B, Carl M (2009) Ten years of translog. Copenhagen Studies in
   Language 37:37–51
Trautner-Kromann M (2003) The Danish dependency treebank and the DTAG tree-
   bank tool. In: 2nd Workshop on Treebanks and Linguistic Theories, Växjo, Swe-
   den, pp 217–220

Creating Linked Data for the Interdisciplinary International Collaborative Study of Language Acquisition and Use: Achievements and Challenges of a New Virtual Linguistics Lab

María Blume, Suzanne Flynn, and Barbara Lust

Abstract In this paper, we describe and exemplify our development of a cyber-tool, the Data Transcription and Analysis tool (DTA tool) that is currently being implemented in the Virtual Center for Language Acquisition through a Virtual Linguistic Lab (VLL). We review this cyber-tool's design and accomplishments to date, assessing its ability to address "the challenge of our time to store, interlink and exploit this wealth of data" (Chiarcos et al., this vol., p. 1). We explicate the architecture and usability of the DTA tool, we summarize its current status, possibilities for expansion, and related challenges we currently confront. We focus on the conceptual and functional structure of this tool here, and not on technical aspects of its programming.

1 Introduction

Data collected from the fields of language acquisition and use are multi-lingual, multi-modal, multi-formatted, and derive from multiple methods of data collection (i.e., observational and experimental, cross-sectional or longitudinal). In addition, they involve multiple aspects of data provenance (e.g., age and/or developmental or cognitive stage of speaker, social and pragmatic contexts, culture). These features result in an immensely complex set of databases often appearing in diverse

María Blume
Department of Languages and Linguistics, University of Texas at El Paso, Liberal Arts Bld. Room 119, El Paso, TX, USA 79968, e-mail: mblume@utep.edu

Suzanne Flynn
Department of Linguistics and Philosophy, MIT, 77 Massachusetts Avenue, 32-DB808, Cambridge, MA 02139, USA, e-mail: sflynn@mit.edu

Barbara Lust
Department of Human Development, Cornell University, G57 Martha Van Rensselaer Hall, Ithaca, NY 14853, USA, e-mail: bcl4@cornell.edu

C. Chiarcos et al. (eds.), *Linked Data in Linguistics*,
DOI 10.1007/978-3-642-28249-2_9, © Springer-Verlag Berlin Heidelberg 2012

formats as different labs generally practice distinct forms of data management. Data from more than 20 languages and cultures and thousands of child and adult subjects exist in the Cornell Language Acquisition Lab and Virtual Center for Language Acquisition alone. The scientific use of any single record requires access to many levels of data, ranging from raw (establishing provenance) to structured and analyzed data (establishing intellectual worth). Language data collections are infinitely expandable and should be used, reused and, when possible, repurposed.

This scientific enterprise requires that these data be stored, accessible and exploited in a manner where relationships can be discovered within and across data sets. In keeping with fundamental insights of the "Linked Data" program (Berners-Lee, 2009), the more each data singleton can be significantly connected or "interlinked", the more powerful and useful it becomes. For example, in the study of language acquisition and use, such interlinks can be cross-disciplinary (e.g., connecting brain images with behavioral experimental results testing language comprehension or production), or linguistically specific (e.g., comparisons of certain properties of sentence structure or verb morphology in a Spanish- versus an English-, French- or Sinhala-speaking child's speech at a particular stage of language development). Data from any one language must be comparable to that in another if one pursues a hypothesis concerning linguistic universals or variation linked to language typology.

2 Approach

In the Virtual Center for Language Acquisition (VCLA),[1] faculty from eight US universities and one international university[2] (Peru) converged through US National Science Foundation, Cornell University and University of Texas at El Paso support to "Transform the Primary Research Process" in the area of language acquisition.[3] A Virtual Linguistics Lab (VLL)[4] was constructed to provide an infrastructure of principles, best practices, materials and cyber-tools including the Data Transcription and Analysis tool.

The Data Transcription and Analysis tool (DTA tool)[5] provides a web-based interface to guide the user – either researcher or student learning scientific methods

---

[1] http://vcla.clal.cornell.edu

[2] Founding members: Suzanne Flynn, MIT; ClaireFoley, Boston College; Liliana Sánchez, Rutgers University, New Brunswick; Jennifer Austin, Rutgers University, Newark; YuChin Chien,California State University at San Bernardino; Usha Lakshmanan, S. Illinois University at Carbondale; Barbara Lust and James Gair, Cornell University; María Blume, University of Texas at El Paso; and Affiliate member Jorge Iván Pérez Silva, Pontificia Universidad Católica del Perú.

[3] Lust, B. 2003 (NSF BCS-0126546); McCue and Lust 2004 (NSF 0437603). Blume and Lust 2007 (NSF OCI-0753415). Seed grant support from the following was also essential to this project: American Institute for Sri Lankan Studies, Cornell University Einaudi Center, Cornell University Faculty Innovation in Teaching Awards, Cornell Institute for Social and Economic Research (CISER), New York State Hatch grant.

[4] http://clal.cornell.edu/vll

[5] http://webdta.clal.cornell.edu

of research – in primary data creation, data management and collaborative data use. At the same time, data entry through this tool automatically feeds a structured, calibrated, and infinitely expandable cross-linguistic relational database. It provides means for structuring, storing and linking scientifically sound, diverse, but calibrated, language data, either naturalistic or experimentally derived, ranging from raw to structured forms, which can be accessed, connected and queried in linked fashion (Lust et al., 2010).

3 Overall Architecture of the DTA Tool

The DTA tool[6] offers the user a structured annotation scheme for the representation of layers of metadata related to language data (i.e., the actual utterances) as well as for representation of reliability-checked transcriptions and analyses of the utterances themselves. Figure 1 provides an overview of the tool's structure showing the major areas from a user's perspective.

The DTA tool is based on 10 tables with the following basic markup categories: Project, dataset, subject, session, recording,[7] transcription, utterance, coding set, coding, and utterance coding.[8] Metadata codings involve the project and subject levels (Fig. 2) and the datasets themselves (Fig. 3) leading to transcribed utterances and related linguistic codings.

The DTA tool also provides some non-project-specific linguistic coding sets: an utterance level coding set (including a literal gloss and a general gloss as well as pragmatic context specification), a set (including speech act and speech mode), and a basic linguistic coding set (including sentence codings and syllable, morpheme and word counts). Users working on natural speech data are expected to use these basic codings, so that the data are calibrated across projects. A researcher working on an experimental project may select if s/he wants to use all, none or some of the established codings. Regardless of the project type, researchers can create new project-specific coding sets or global coding sets.[9]

---

[6] We concentrate on the latest 2011 version of the DTA tool, which has been reprogrammed by GORGES, a web and mobile technology firm in Ithaca, New York. (http://www.gorges.us). The DTA tool has been under construction for more than 20 years through several generations of students who contributed to its development and through several changing technical formats. For a history of its development see the DTA User's Manual (Blume and Lust, 2011; Blume et al., in prep).
The current version of the WebDTA tool is built on Yii, a PHP web development framework that uses the 'Model-View-Controller' pattern to structure the application and the 'ActiveRecord' pattern to manage records from the database. MySQL is used for the database platform. All are open source technologies.

[7] Digital audio or video file, an electronic document (e.g. a Word, Excel, or PDF file), or information in a non-digital format such as a tape recording or paper transcription.

[8] An utterance coding records specified linguistic values (which the DTA tool refers to as 'codings') for a given utterance. 'Subject' refers to the participant providing language to be studied.

[9] Selected permission level is necessary for such new coding creation.

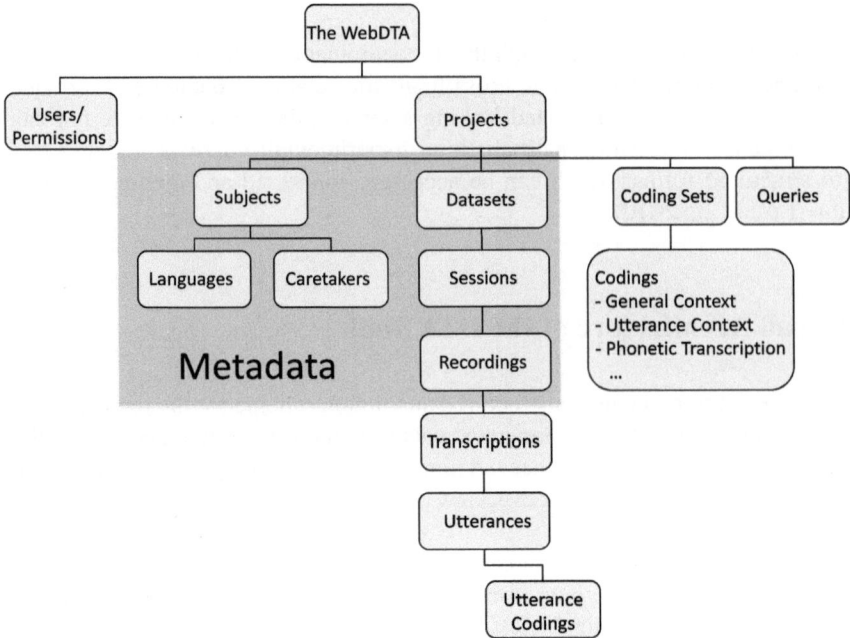

Fig. 1 WebDTA tool basic structure

Figure 4 exemplifies basic coding of an utterance of natural speech data of a Peruvian monolingual Spanish-speaking child from the 'Spanish Natural Speech-Blume' corpus. Such codings render the data ready for further analyses in connection with specific research questions.

4 Examples of DTA Tool Content

Users complete a series of tables through a graphical user interface intended to facilitate and structure a user's data and metadata entry. The tool first leads the user to enter detailed project-level information and metadata. Main areas include project investigators, purpose and leading hypotheses, references (publications, presentations, related studies, bibliography), subjects, and results and discussion. At several different points documents can be attached. The tool also provides summary reports showing subsets of metadata in addition to the datasets of the project.

The user enters information on the datasets that form a project. For each dataset, the user provides the main information (experiment/investigation, topic, abstract, related WebDTA projects/datasets), hypotheses, general subject description, methods, design, stimuli, procedures, and scoring procedures. When a research project is completed, results and conclusions can be linked. The DTA tool provides a report at

Project

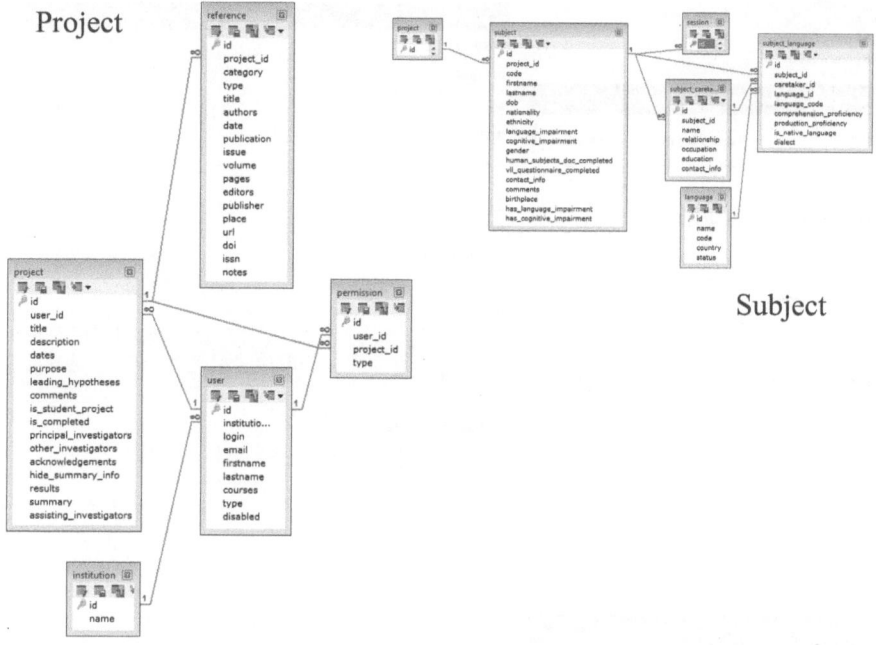

Fig. 2 Project and subject metadata

Subject

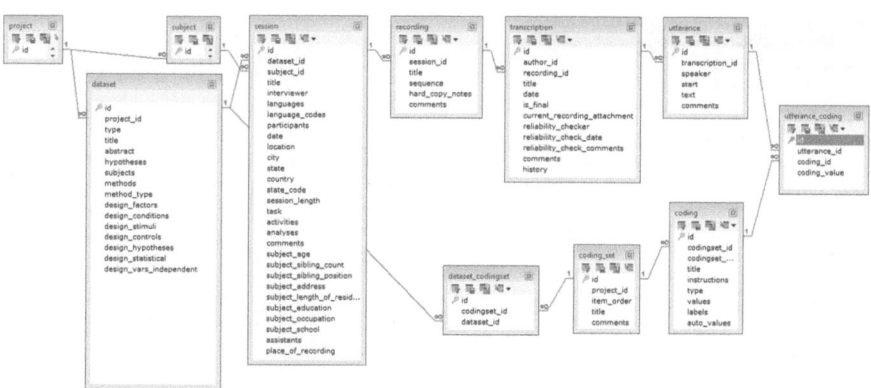

Fig. 3 Dataset metadata

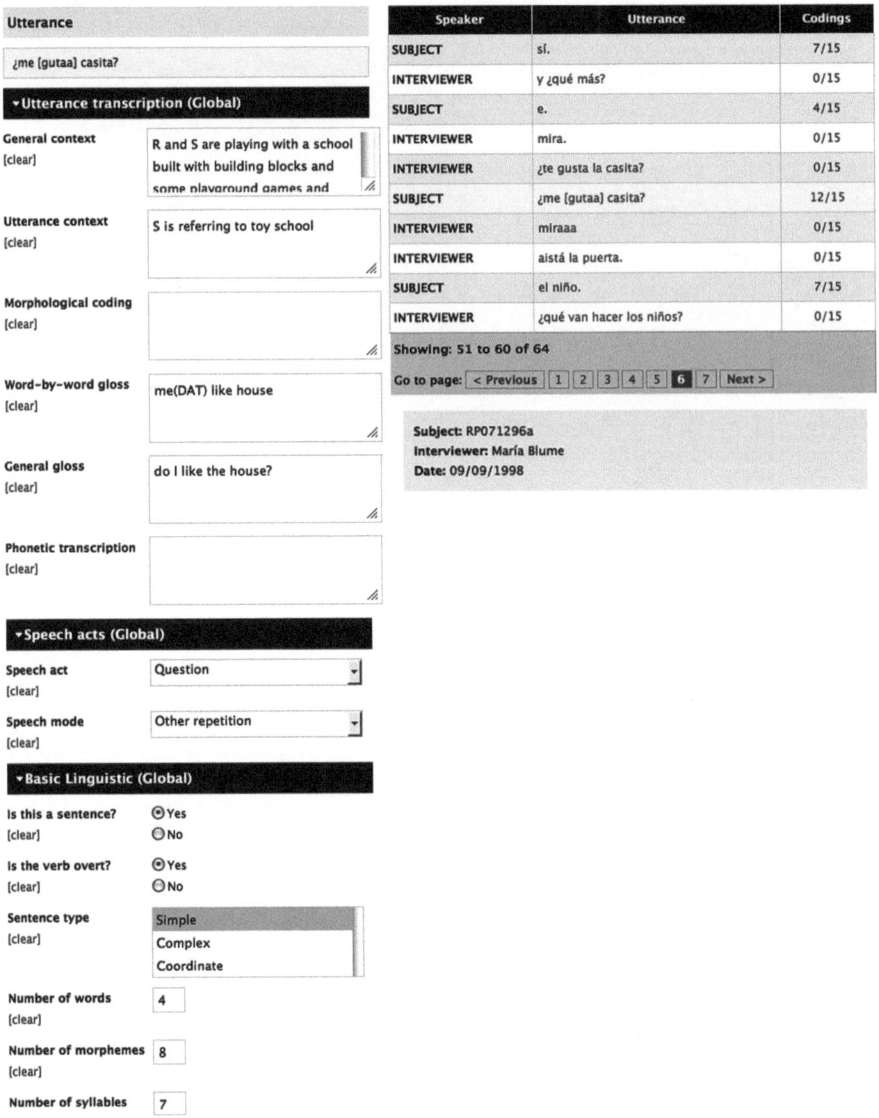

Utterance		
¿me [gutaa] casita?		

▼Utterance transcription (Global)

General context [clear]	R and S are playing with a school built with building blocks and some playground games and

Utterance context [clear]	S is referring to toy school

Morphological coding [clear]	

Word-by-word gloss [clear]	me(DAT) like house

General gloss [clear]	do I like the house?

Phonetic transcription [clear]	

▼Speech acts (Global)

Speech act [clear]	Question
Speech mode [clear]	Other repetition

▼Basic Linguistic (Global)

Is this a sentence? [clear]	⊙Yes ○No
Is the verb overt? [clear]	⊙Yes ○No
Sentence type [clear]	Simple / Complex / Coordinate
Number of words [clear]	4
Number of morphemes [clear]	8
Number of syllables	7

Speaker	Utterance	Codings
SUBJECT	sí.	7/15
INTERVIEWER	y ¿qué más?	0/15
SUBJECT	e.	4/15
INTERVIEWER	mira.	0/15
INTERVIEWER	¿te gusta la casita?	0/15
SUBJECT	¿me [gutaa] casita?	12/15
INTERVIEWER	miraaa	0/15
INTERVIEWER	aistá la puerta.	0/15
SUBJECT	el niño.	7/15
INTERVIEWER	¿qué van hacer los niños?	0/15

Showing: 51 to 60 of 64

Go to page: < Previous 1 2 3 4 5 **6** 7 Next >

Subject: RP071296a
Interviewer: María Blume
Date: 09/09/1998

Fig. 4 Coding screen: Example of Spanish child language utterance entry

the dataset level, the Experiment Bank report, including all the information for the project and each of its datasets.

After the user provides specified metadata for all the subjects in the project, the user enters information for the sessions pertaining to each dataset. Each session[10] has associated to it a recordings screen, a transcription screen, and a coding screen.

---

[10] A 'session' refers to a particular time in which a particular set of language data is recorded.

The recordings screen houses information on all available primary data for a given session (audio or video files or previous transcripts in a number of formats) plus an inventory of the location of such files, and their backups. This screen supports all files supported by the JW Player,[11] *QuickTime* player, PDF, HTML, and image files, and, with additional software, other file formats such as *Microsoft Office* files. The user then moves to a transcription screen where he/she can watch and listen to all available recordings (switching from one to the other as needed), transcribe and manually set timings to align the transcript with the recordings. Finally, the user moves to the coding screen where he/she can code for any of the global coding sets or for codings created specifically for the particular project. Figure 5 illustrates a specific coding set created for an utterance from a Peruvian child participant in the experimental Project, "Discourse Morphosyntax Interface in Spanish Non-Finite Verbs-Blume".

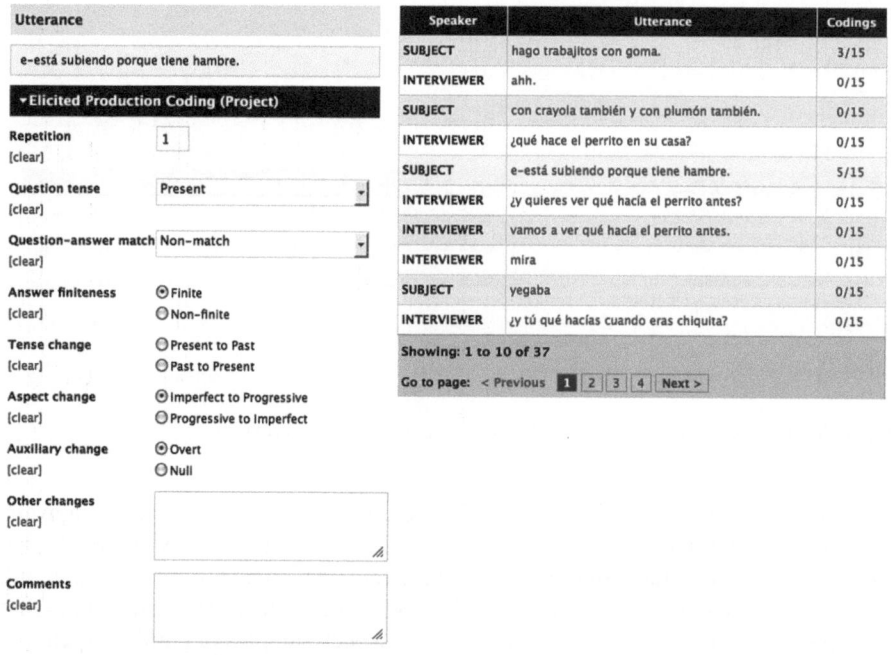

Fig. 5 Project-specific linguistic coding set

---

[11] JW Player: H.264 video (.mp4, .mov, .f4v), Flash video (.flv), YouTube video, 3GPP video (.3gp, .3g2), MP3 audio (.mp3), and AAC audio (.aac, .m4a). QuickTime: AIFF audio (.aif, .aiff). Image files: .jpg, .jpeg, .gif, .png

5 Linking Data Through Queries

A set of queries, which is essential to calibrating language data, is available initially in the DTA tool. Queries search properties of the data such as MLU (Mean Length of Utterance) in words, syllables, and morphemes; coded information on all subjects across projects who speak a particular language or have a certain age, utterances that are sentences, utterances that are NPs, sentences with overt verbs, simple vs. complex sentences, specific speech acts or speech modes. Figure 6 illustrates a basic query searching for all simple sentences with verbs for a particular session for a subset of one subject's data in the 'Spanish Natural Speech Corpus-Blume' corpus.[12] Figure 7 shows the results of the query. Queries can be run on all sessions that have been coded for the relevant features in all projects in the DTA tool, thus linking across sessions and subjects.

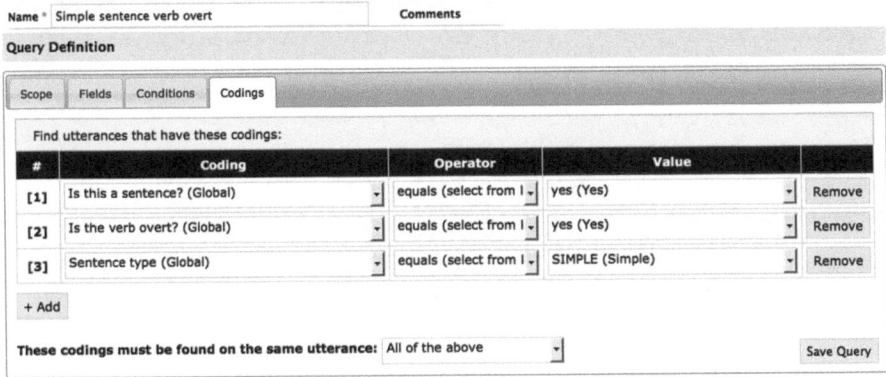

Fig. 6 Values of codings in query for all simple sentences with overt verbs.

In addition, the tool allows continual generation of new queries based on available codings that derive from a particular research question. For example, we can ask about a specific matching of tense and aspect forms in a verb in a question and in the answer in an Elicited Production task. We can also query about sentences that exemplify standard word orders across natural speech samples of children speaking different languages at the same MLU stage, etc.

Given the possibility for such content generation through the DTA tool, it can be seen that the DTA tool is a *primary* research tool, guiding the researcher or student in data collection and management; its potential usability extends from specific research projects to use in educational domains as well. At the same time, the tool automatically provides a rich, continually growing archive allowing present and future collaboration on shared data, potentially long distance, and potentially interdisci-

---

[12] To create a query the user would also need to define the scope, conditions for the query and the fields one would like to see when results are found, which we do not show here due to space limitations.

Query Results (15 records)

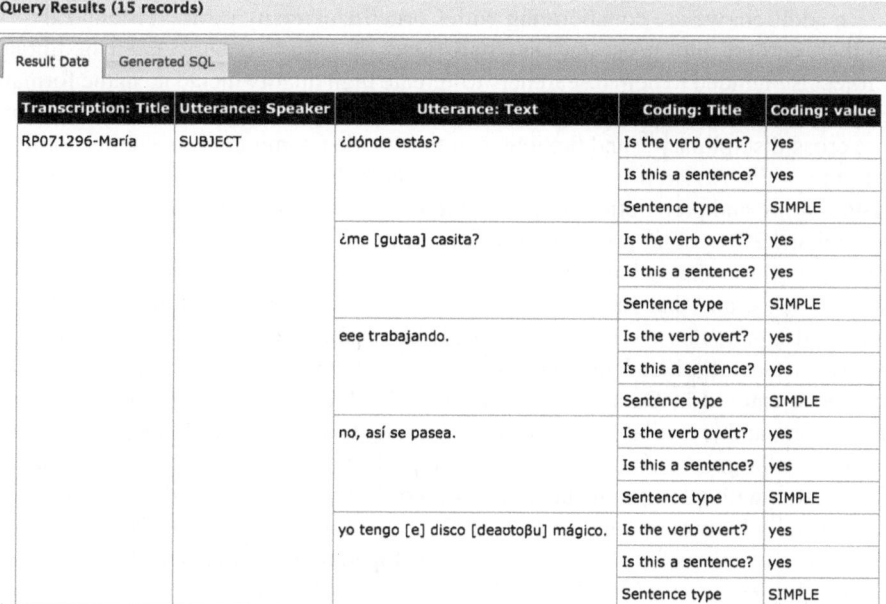

Transcription: Title	Utterance: Speaker	Utterance: Text	Coding: Title	Coding: value
RP071296-María	SUBJECT	¿dónde estás?	Is the verb overt?	yes
			Is this a sentence?	yes
			Sentence type	SIMPLE
		¿me [gutaa] casita?	Is the verb overt?	yes
			Is this a sentence?	yes
			Sentence type	SIMPLE
		eee trabajando.	Is the verb overt?	yes
			Is this a sentence?	yes
			Sentence type	SIMPLE
		no, así se pasea.	Is the verb overt?	yes
			Is this a sentence?	yes
			Sentence type	SIMPLE
		yo tengo [e] disco [deaʊtoβu] mágico.	Is the verb overt?	yes
			Is this a sentence?	yes
			Sentence type	SIMPLE

Fig. 7 Previous query's results

plinary.[13] In general, with external linkages, through Linked Data formats, the DTA tool's database can be linked to a wide intellectual knowledge base, e.g., linking published forms of research to the actual data and data methods used to create the results reported.

6 Further Linkages: External

The DTA tool is designed to maximize the possibility for Linked Data by integrating with field standards. For example, the application uses the UTF-8 encoding to store text, which can represent any language.[14] For this, the application adopts ISO 639-3 standard language codes, which lists over 7000 languages, developed by Ethnologue/SIL.[15] It links GeoNames.org in geographic reference.

---

[13] Examples of collaborative research projects including students and researchers are at Cornell (Barbara Lust): "SAQL Phase 1: Expert Evaluation and Validation of a New Child Multilingualism Questionnaire", Newcastle University (Cristina Dye) and Boston College (Claire Foley): "Acquisition of VP ellipsis in mono and bilingual children"; MIT (Suzanne Flynn), Massachusetts General Hospital and Cornell (Barbara Lust): "Alzheimer's language project".

[14] Given the availability of language specific fonts.

[15] http://www.ethnologue.com/codes/default.asp

In addition, we are collaborating with Cornell University's Albert Mann Library in their current pilot program, DataStaR (Data Staging Repository).[16] This library project is intended to help researchers to "create high quality metadata in the formats required by external repositories..." (Steinhart, 2010, p. 1) where metadata systems are cross-disciplinary[17] and flexible, and to provide a temporary repository for data sharing while research is in progress. DataStaR thus fosters an infrastructure for data sharing and preparation for publication to external repositories where available (Steinhart, 2010). The program adopts a semantic web approach to metadata "on the assumption that scientific communities will increasingly adopt semantic web technologies, and that Linked Data will become increasingly common..." (Steinhart, 2010, p. 5; also see Khan et al., 2011; Lowe, 2009, for details). At present, one VCLA dataset (Sinhala language) from more than 400 children studied in Sri Lanka has been entered in DataStaR, linking the VCLA database to the Library staging repository, and is available for collaborative use through this repository.[18]

As our linguistic coding system develops further, we intend to pursue further integration with attempts at standardization of linguistic category descriptions such as GOLD[19] (Farrar and Langendoen, 2003), also see (Simons et al., 2004); as well as with conventions for glossing such as developed by Bickel et al. (2008). We have initiated integration with OLAC (Open Language Archives Community) and are pursuing further development of this link.[20]

7 Current Status and Challenges

At this point we are completing beta-testing of the current newly programmed version of the DTA tool (cf. fn 5). This new version of the tool is currently being populated by data from a previous version of the tool that is now being ported to this new version, and by new data from several current collaborative research and educational projects.

Several cross institutional courses, including two with Peru, at the undergraduate and graduate levels have introduced the VLL and the DTA tool through our structured VLL web portal[21] and a series of web conferences over the last three years in

---

[16] Funded by the National Science Foundation (Grant No. 111-0712989)

[17] DataStaR uses RDF (Resource Description Framework (RDF)) statements and OWL (Web Ontology Language) classes in order to integrate different metadata frameworks across disciplines.

[18] http://datastar.mannlib.cornell.edu/display/n6291 and http://www.news.cornell.edu/stories/Oct11/SinhalaTools.html

[19] Generalized Ontology for Linguistic Description. 2010. Available online at http://www.linguistics-ontology.org/gold.htm

[20] Open Language Archives Community (OLAC), http://www.language-archives.org/ (24 Feb. 2011) (See also Open Archives Initiative (OAI), http://www.openarchives.org/ (15 Mar. 2005)).

[21] The VLL portal is accessible under http://www.clal.cornell.edu/vll. It was constructed by Tommy Cusick, previous Cornell student, now Google, Inc.

order to train a new generation of scholars in scientific methods for data collection and management.

The main challenges we must address now include the following: (i) In order to open the DTA tool productively to a wide audience, we must build a sustainability model that includes licensing options. For this we have now initiated correspondence with Cornell's E-Cornell program (eCornell.com); (ii) To widely extend the DTA tool to new users we must establish a set of and agreements involving shared materials and data. This must involve establishment of a leveled set of permissions, e.g., read only, etc. Founding Members of the VCLA are currently addressing this challenge; (iii) To ensure long-term sustainability, we must negotiate long-term storage of the database; (iv) We must develop an infrastructure for long-term management of the tool and its access and use. In our view this would ideally be some form of a distributed infrastructure rather than a localized one.

In addition, several technical challenges remain as the new version of the DTA tool is being activated. For example, porting data from an old version of the DTA tool to the new directly instantiates the general issue of database linkage. We are currently working with Cornell Information Technologies to pursue the possibility for alternative video and audio streaming. A version of the tool allowing asynchronous uploading of data should be developed to facilitate work in field situations, e.g., cross-linguistic work, without Internet access.

Acknowledgements Several grants have provided funding for this project. The main support for our current development comes from Blume, María and Barbara Lust. 2007. Transforming the Primary Research Process Through Cybertool Dissemination: An Implementation of a Virtual Center for the Study of Language Acquisition. National Science Foundation. Award No. OCI-0753415. Other grants that have provided with fundamental resources are: Lust, Barbara. 2003. Planning Grant: A Virtual Center for Child Language Acquisition Research. National Science Foundation. NSF BCS-0126546, and Janet McCue and Barbara Lust 2004-2006. National Science Foundation Award: Planning Information Infrastructure Through a New Library-Research Partnership. (SGER=Small Grant for Exploratory Research). Seed grant support from the following was also essential to this project: American Institute for Sri Lankan Studies, Cornell University Einaudi Center, Cornell University Faculty Innovation in Teaching Awards, Cornell Institute for Social and Economic Research (CISER), New York State Hatch grant. We would also like to acknowledge the contributions to this project of VCLA members and affiliates: Elise Temple (NeuroLab) Gita Martohardjono (CUNY), Cristina Dye (Newcastle University), Isabelle Barrière (CUNY), Yarden Kedar (Ben Gurion University at the Negev), and Sujin Yang (Tyndale University College and Seminary); of our application developers Ted Caldwell and Greg Kops (GORGES); of our consultants Cliff Crawford and Tommy Cusick; of our student RAs Darlin Alberto, Gabriel Clandorf, Natalia Buitrago, Poornima Guna, Jennie Lin, and Jordan Whitlock at Cornell, and Marina Kalashnikova. Martha Rayas Tanaka, Lizzeth Pattison, María Jiménez, and Mónica Martínez at UTEP. We would also liked to thank the students at all the participating institutions that helped us with comments and suggestions.

References

Berners-Lee T (2009). URL http://en.wikipedia.org/wiki/Linked_data, Ted Lecture. Tim Berners-Lee on the next Web

Bickel B, Comrie B, Haspelmath M (2008) Glossing rules: Conventions for inter-linear morpheme-by-morpheme glosses. URL http://www.eva.mpg.de/lingua/resources/glossing-rules.php

Blume M, Lust B (2011) Data Transcription and Analysis Tool User's Manual. With the collaboration of Shamitha Somashekar, and Tina Ogden

Blume M, Yang S, Lust B (in prep) Virtual Linguistics Lab (VLL) Research Methods Manual: Scientific Methods for Study of Language Acquisition. With the collaboration of Tina Ogden, Shamitha Somashekar, Yuchin Chien, Liliana Sánchez, Claire Foley, Marina Kalashnikova, Martha Rayas, Yarden Kedar and Natalia Buitrago

Chiarcos C, Hellmann S, Nordhoff S (this vol.) Introduction and overview. pp 1–12

Farrar S, Langendoen D (2003) A linguistic ontology for the semantic web. GLOT International 7:97–100

Khan H, Caruso B, Lowe B, Corson-Rikert J, Dietrich D, Steinhart G (2011) Datastar: Using the semantic web approach for data curation. International Journal of Digital Curation 2:209–221

Lowe B (2009) Datastar: Bridging XML and OWL in science metadata management. Metadata and Semantics Research 46:141–150, URL http://www.springerlink.com/content/q0825vj78ul38712/

Lust B, Flynn S, Blume M, Westbrooks E, Tobin T (2010) Constructing adequate documentation for multi-faceted cross linguistic language data: A case study from a virtual center for study of language acquisition. In: Language Documentation: Theory, Practice and Values, John Benjamins, Amsterdam/Philadelphia, pp 127–152

Simons G, Lewis W, Farrar S, Langendoen D, Fitzsimons B, Gonzalez H (2004) The semantics of markup. In: Proc. 4th Workshop on NLP and XML (NLPXML-2004), Barcelona, Spain, pp 25–32

Steinhart G (2010) Datastar: A data staging repository to support the sharing and publication of research data. In: 31st Annual IATUL Conference - The Evolving World of e-Science: Impact and Implicationsfor Science and Technology Libraries, West Lafayette, URL http://docs.lib.purdue.edu/iatul2010/conf/day2/8/

Part III
Terminology Repositories and Knowledge Bases

Linking to Linguistic Data Categories in ISOcat

Menzo Windhouwer and Sue Ellen Wright

Abstract ISO Technical Committee 37, Terminology and other language and content resources, established an ISO 12620:2009 based Data Category Registry (DCR), called ISOcat (see http://www.isocat.org), to foster semantic interoperability of linguistic resources. However, this goal can only be met if the data categories are reused by a wide variety of linguistic resource types. A resource indicates its usage of data categories by linking to them. The small DC Reference XML vocabulary is used to embed links to data categories in XML documents. The link is established by an URI, which servers as the Persistent IDentifier (PID) of a data category. This paper discusses the efforts to mimic the same approach for RDF-based resources. It also introduces the RDF quad store based Relation Registry RELcat, which enables ontological relationships between data categories not supported by ISOcat and thus adds an extra level of linguistic knowledge.

1 Introduction

ISO Technical Committee 37 Terminology and other language and content resources established a Data Category Registry (DCR), called ISOcat, to foster semantic interoperability of linguistic resources. ISOcat is based on ISO 12620:2009, which describes the data model and the management procedure for a DCR (ISO 12620, 2009). These procedures follow a grass roots approach, which means that any linguist can add the data categories (s)he needs to the registry. Standardized subsets of these data categories are created by a standardization procedure involving groups of

Menzo Windhouwer

Max Planck Institute for Psycholinguistics, Wundtlaan 1, 6525 XD Nijmegen, The Netherlands,
e-mail: Menzo.Windhouwer@mpi.nl

Sue Ellen Wright

Kent State University, 109 Satterfield Hall, Kent, OH 44242, USA e-mail: sellenwright@gmail.com

C. Chiarcos et al. (eds.), *Linked Data in Linguistics*,
DOI 10.1007/978-3-642-28249-2_10, © Springer-Verlag Berlin Heidelberg 2012

international experts who are members of various Thematic Domain Groups (TDGs) and the DCR Board. There are currently over a dozen domains supported by a TDG, e.g., metadata, morphosyntax and terminology. But the aim of improving the semantic interoperability can only be met by the data categories if they are reused by a multitude of linguistic resource types (Kemps-Snijders et al., 2008). A resource indicates its usage of data categories by linking to them (Windhouwer et al., 2010). This paper focuses on how this can be done, and gives special attention to linked open data, i.e., RDF-based, resources.

2 Linking to Data Categories from XML-Based Resources

The focus of ISOcat has been mainly on general XML-based resources. ISO 12620:2009 specifies a small Data Category (DC) Reference XML vocabulary (see http://www.isocat.org/12620/) to annotate XML documents with links to data categories. This vocabulary, using Relax NG compact syntax as the schema language, defines two basic annotation descriptors, as shown in Example 1.[1]

```
default namespace dcr = "http://www.isocat.org/ns/dcr"

dcr_attribute_datcat = attribute datcat { xsd:anyURI }
dcr_attribute_value_datcat = attribute valueDatcat { xsd:anyURI }
```

Example 1: DC Reference attributes specified in Relax NG

The dcr:datcat descriptor can be used to annotate any XML element with a link to the equivalent data category in ISOcat. The other descriptor, dcr:valueDatcat, is in general used to annotate the textual value of an element or attribute, i.e., to annotate this value with a link to a simple data category.

2.1 Persistent Identifiers

As Example 1 shows, the link to the data category in ISOcat should be established by a URI. In ISOcat each data category has a unique and persistent identifier, also known as the PID (Persistent Identifier). The URI scheme that ISOcat uses for the PID is called a 'cool URI' (Berners-Lee, 1998), which is basically a standard HTTP URL with extra guarantees that these URLs will remain resolvable over a long period of time. The use of cool URIs is only one of the possible approaches to creating

---

[1] The DC Reference XML vocabulary defines the descriptors both as XML attributes and XML elements. The specific structure of the annotated XML-based resource determines whether either the attribute or the element should be used.

PIDs. ISO TC 37 has recently published a new standard, PISA (Persistent Identification and Sustainable Access, ISO 14619:2011), which describes the requirements to be met by these PID systems.

2.2 Data Category Types

In the structure of a resource various elements play different roles, e.g., some elements can have values while other elements group other elements. To accommodate these various roles there are data categories of different types:

1. Complex data categories have a typed value domain; the DCR data model supports various ways to describe these value domains:
   a. Open data categories can take any value allowed by the associated type;
   b. Closed data categories enumerate their allowed values as simple data categories (see below);
   c. Constrained data categories restrict their allowed values by one or more rules, e.g., any day in the 20th century;
2. Simple data categories are describe values associated with a closed data category;
3. Container data categories don't have a value domain but can be used to group other container or complex data categories together.

ISOcat does not store any relationships beyond the basic value domain relationships between simple and closed data categories. This means that specific structures built using container and complex data categories are not available in ISOcat as this would hamper their reuse. These structures are preferably specified in a resource schema document, e.g., a W3C XML Schema or Relax NG document, annotated with data category references.

2.3 An Annotated LMF Document

The ISO standard for the Lexical Markup Framework, ISO 24613:2008, encourages the use of ISOcat data categories although it omits to mention the DC Reference XML vocabulary. Example 2, taken from the standard (ISO 24613, 2008), has been annotated with ISOcat data category PIDs.[2]

It is apparent that the annotation of an instance quickly becomes verbose. A general solution is to annotate the schema of the resource, e.g., the W3C XML Schema or Relax NG schema. This way many instances can be annotated in using a single resource.

---

[2] Due to space limitations the common ISOcat cool URI prefix `http://www.isocat.org/datcat` has been replaced by elipses.

```
<LexicalResource xmlns:dcr="http://www.isocat.org/ns/dcr">
  <GlobalInformation>
    <feat att="languageCoding" dcr:datcat=".../DC-2008"
      val="ISO 639-3"/>
  </GlobalInformation>
  <Lexicon>
    <feat att="language" dcr:datcat=".../DC-1969" val="eng"/>
    <LexicalEntry>
      <feat att="partOfSpeech" dcr:datcat=".../DC-1345"
        val="commonNoun" dcr:valueDatcat=".../DC-1256"/>
      <Lemma>
        <feat att="writtenForm" dcr:datcat=".../DC-1836"
          val="clergyman"/>
      </Lemma>
      ...
      <WordForm>
        <feat att="writtenForm" dcr:datcat=".../DC-1836"
          val="clergymen"/>
        <feat att="grammaticalNumber" dcr:datcat=".../DC-1298"
          val="plural" dcr:valueDatcat=".../DC-1354"/>
      </WordForm>
    </LexicalEntry>
  </Lexicon>
</LexicalResource>
```

Example 2: Annotated LMF example

The example doesn't show the use of container data categories as this is a recent addition to the DCR data model not even covered by ISO 12620:2009. For the LMF core model and its extensions these container data categories have not been specified yet. However, it does show an open data category, i.e., /writtenForm/ (http://www.isocat.org/datcat/DC-1836), an simple data category, i.e., /commonNoun/ (http://www.isocat.org/datcat/DC-1256) which is an instance of the value domain from a closed data category, i.e., /partOfSpeech/ (http://www.isocat.org/datcat/1345).

3 Linking to Data Categories from RDF-Based Resources

Using the descriptors described above any XML document can refer to data categories to make the semantics explicit for elements, attributes and values. However, determining the level of semantic interoperability still involves additional processing, e.g., determining the overlap in semantics by finding the shared data categories. In the Linked Data world of RDF resources this interoperability is built into the data model used, i.e., data categories can play a direct role in that model as they are resources identifiable by a (cool) URI.

3.1 RDF Resource and Data Category Types

At first glance there seems to be a natural correspondence between data category types and RDF types: container data categories correspond to RDF classes, complex data categories correspond to RDF properties, and simple data categories correspond to literal values. However, this simple mapping comes with some drawbacks:

- Data categories would be far more prominent in the RDF model than in the XML model, as the direct use of the data categories would impose their non-semantics bearing URIs on the model being constructed. In the case of an XML model the fact that data categories are references leaves the choice of terminology more to the model builder. This is also more in line with the terminology management background of the DCR, which is also reflected by the data model which allows the specification of various (technical) terms used for a data category in a variety of circumstances, e.g., language or application specific.
- Literal values cannot be annotated in an RDF model, which means that the appropriate simple data category cannot be referred to when the literal value is ambiguous in the profile value domains of the closed data category. One possible solution could be to use simple data categories as individuals, but in this case the Cool URL of simple data category would again feature prominently in the RDF model.
- RDF models can actually be used to fine tune the ontological relationships between data categories, and specifying fixed type would hinder this as these relationships might require different mappings to RDF types.

For now ISOcat leaves the actual mapping to RDF types to the model builder and outputs the data categories as related to RDF resources. The model builder can then decide which types are appropriate. Also the data category cool URIs are, just like in the XML world, used to annotate these RDF resources so the model builder can fine tune these resources, e.g., using his/her own terminology. Experiments with approaches to mapping will continue, and either simple or more advanced forms of mapping might even be used for a (semi-)automatic conversion to RDF for annotated (either inline or by their schema) XML documents.

Example 3 shows an RDF specification for a simple model for a dictionary annotated with data category references.

3.2 RDF Predicates and Data Category Links

The example also shows the use of the `dcr:datcat` predicate to associate the RDF resource with the data category (Example 4).

The first version of the ISOcat RDF export used `owl:sameAs`. But the drawback of that approach is that by using OWL semantics the annotation has impact on the OWL model being built, i.e., it quickly pushes the OWL model to OWL Full.

```
@prefix dcr: <http://isocat.org/ns/dcr.rdf#> .

:headword dcr:datcat <http://isocat.org/datcat/DC-258> ;
  rdfs:label "head word"@en ;
  rdfs:comment "A lemma heading a dictionary entry."@en ;
...
:partOfSpeech dcr:datcat <http://isocat.org/datcat/DC-396> ;
  rdfs:label "part of speech"@en ;
  rdfs:comment "A category assigned to a word based on its
               grammatical and semantic properties."@en .
...
```

Example 3: Annotated RDF resource

```
@prefix dcr: <http://isocat.org/ns/dcr.rdf#> .

dcr:datcat a owl:AnnotationProperty ;
  rdfs:label "data category"@en ;
  rdfs:comment "This resource is equivalent to this data
    category."@en ;
  skos:note "The data category should be identified by
    its Persistent IDentifier (PID)."@en ;
...
```

Example 4: dcr:datcat annotation property

This is an unwanted side effect and is prevented by specifying a dedicated annotation property. Once more the RDF model builder can fine tune this. Depending on the actual RDF type of the annotated RDF resource the dcr:datcat predicate can be replaced by the following OWL (2) predicates: owl:equivalentClass for classes, owl:equivalentProperty for properties and owl:sameAs for individuals. The use of these specific predicates limits the impact of ISOcat data categories on OWL semantics.

4 Ontological Relationships

ISOcat basically contains a flat list of data categories, i.e., it doesn't store (ontological) relationships between container and/or complex data categories. In addition to value domain relationships between simple and closed data categories, only a subsumption hierarchy between simple data categories is stored, but only one such a subsumption hierarchy is allowed, i.e., a simple data category can only be a child of one other data category. The storage of these ontological relationships in ISOcat is due to legacy issues and its usage is actually discouraged.

The reason that ontological relationships aren't stored in ISOcat is that they are highly domain or even application dependent and thus would hamper standardization of data category specifications. However, they are important to make the semantics of linguistic resources explicit. To support this a companion registry to ISOcat named RELcat is under construction (Schuurman and Windhouwer, 2011). In RELcat anyone or any group can store (ontological) relationships between data categories and/or concepts from other registries.

```
@prefix relcat    : <http://www.isocat.org/relcat/set/> .
@prefix rel       : <http://www.isocat.org/relcat/relations#> .
@prefix dc        : <http://purl.org/dc/elements/1.1/> .
@prefix isocat    : <http://www.isocat.org/datcat/> .

relcat:cmdi {
    isocat:DC-2573 rel:sameAs dc:identifier .
    isocat:DC-2482 rel:sameAs dc:language .
    ...
    isocat:DC-2556 rel:subClassOf dc:contributor .
    isocat:DC-2502 rel:subClassOf dc:coverage .
}
```

Example 5: Relations between data categories and Dublin Core elements

Example 5 shows the set of relationships between data categories in ISOcat and Dublin Core elements. This set is in use by the metadata search engine for the CLARIN MetaData Infratructure (Broeder et al., 2008, CMDI). Mappings to support crosswalks to any other (linguistic) metadata element set, e.g., OLAC (Simons and Bird, 2003), or ontologies or taxonomies, e.g., GOLD (Farrar and Langendoen, 2010), can be added in the same vein.

4.1 Ontological Relationship Types

RELcat supports the following ontological relationship types:

1. related
    a. same as (a symmetric and transitive relationship)
    b. almost same as (a symmetric relationship)
    c. broader than (a transitive relationship and the inverse of the 'narrower than' relationship)
        i. superclass of (a transitive relationship and the inverse of the 'subclass of' relationship)
        ii. has part (a transitive relationship and the inverse of the 'part of' relationship)
            A. has direct part (the inverse of the 'direct part of' relationship)
    d. narrower than (a transitive relationship and the inverse of the 'broader than' relationship)

    i. sub class of (a transitive relationship and the inverse of the 'super class of' relationship)

    ii. part of (a transitive relationship and the inverse of the 'has part' relationship)

       A. direct part of (the inverse of the 'has direct part' relationship)

Although inspired by OWL and SKOS these relationship types may seem to be an impoverished set. But they are already an extension to the original purpose of RELcat, which mainly dealt with (almost) same-as relationships. However, this shallow taxonomy is just a first start. Other relationship types from other richer vocabularies, e.g., complete OWL or SKOS, can be inserted at the proper place in this subsumption hierarchy:

1. related

    a. same as (a symmetric and transitive relationship)
       i. `owl:equivalentClass`
       ii. `owl:equivalentProperty`
       iii. `owl:sameAs`
       iv. `skos:exactMatch`
    b. almost same as (a symmetric relationship)
       i. `skos:closeMatch`
    c. ...

Now sets of relations using these vocabularies can be loaded into RELcat, and be combined and exploited in their usual fashion, e.g., by an inferencing engine. For example, this is done for the GOLD ontology of linguistic concepts (Farrar and Langendoen, 2010). However, the upper part of the taxonomy can be used by generic algorithms to traverse the large graph created by the combined relationships.

```
PREFIX rel:<http://www.isocat.org/relcat/relations#>
PREFIX isocat:<http://www.isocat.org/datcat/>

SELECT ?rel WHERE { isocat:DC-2482 rel:related ?rel . }
```

Example 6: SPARQL query for relations with */languageID/* (`http://www.isocat.org/datcat/2482/`)

The query in Example 6 returns both the Dublin Core `dc:language` metadata element and the Language (`http://purl.org/linguistics/gold/Language`) GOLD concept, although their relationships with the */languageID/* (`http://www.isocat.org/datcat/2482`) data category have been expressed using different RDF vocabularies. One of the purposes of RELcat is to provide this information to semantic search engines to enable the retrieval of closely related resources of different types.

5 Conclusion and Future Work

Although the ISOcat data model is expressed in a more conventional UML data model the registered data categories can actually easily be used in the context of Linked Data due to the use of cool URIs as PIDs. This paper discussed the use of dedicated annotation attributes and properties to annotate existing XML and RDF documents. It also discussed some if the design decisions for RELcat, a Relation Registry, which enabled to specify ontological relationships among ISOcat data categories but also with concepts from other registries. Future work includes further development of RELcat aiming at achieving a higher level of semantic interoperability.

References

Berners-Lee T (1998) Cool URIs don't change. Tech. rep., World Wide Web Consortium, http://www.w3.org/Provider/Style/URI.html

Broeder D, Declerck T, Hinrichs E, Piperidis S, Romary L, Calzolari N, Wittenburg P (2008) Foundation of a component-based flexible registry for language resources and technology. In: Proceedings of the 6th International Conference on Language Resources and Evaluation (LREC 2008), Marrakech, Morocco

Farrar S, Langendoen DT (2010) An OWL-DL implementation of GOLD: An ontology for the semantic web. In: Witt AW, Metzing D (eds) Linguistic Modeling of Information and Markup Languages: Contributions to Language Technology, Springer

ISO 12620 (2009) Terminology and other language and content resources - Specification of data categories and management of a Data Category Registry for language resources

ISO 24613 (2008) Language resource management - Lexical markup framework (LMF)

Kemps-Snijders M, Windhouwer M, Wittenburg P, Wright SE (2008) ISOcat: Corralling data categories in the wild. In: Proceedings of the Sixth International Conference on Language Resources and Evaluation (LREC'08), Marrakech, Morocco, http://www.lrec-conf.org/proceedings/lrec2008/

Schuurman I, Windhouwer M (2011) Explicit semantics for enriched documents. What do ISOcat, RELcat and SCHEMAcat have to offer? In: Proceedings of the 2nd Supporting Digital Humanities Conference, Copenhagen, Denmark

Simons G, Bird S (2003) The open language archives community: An infrastructure for distributed archiving of language resources. Literary and Linguistic Computing 18(2):117–128

Windhouwer M, Wright SE, Kemps-Snijders M (2010) Referencing ISOcat data categories. In: Budin G, Declerck T, Romary L, Wittenburg P (eds) Proceedings of the LREC 2010 LRT standards workshop, Malta, http://www.lrec-conf.org/proceedings/lrec2010/workshops/W4.pdf

Towards Linked Language Data for Digital Humanities

Thierry Declerck, Piroska Lendvai, Karlheinz Mörth, Gerhard Budin, and
Tamás Váradi

Abstract We investigate the extension of classification schemes in the Humanities
into semantic data repositories, the benefits of which could be the automation of
so far manually conducted processes, such as detecting motifs in folktale texts. In
parallel, we propose linguistic analysis of the textual labels used in these repositories. The resulting resource, which we propose to publish in the Linked Open Data
(LOD) framework, will explicitly interlink domain knowledge and linguistically enriched language data, which can be used for knowledge-driven content analysis of
literary works.

1 Introduction

We discuss strategies of porting semi-structured resources in the field of folk
literature into the expanding linked open data (LOD) framework.[1] Prominent
examples of such resources are the "Thompson Motif-Index of folk-literature"
(Thompson, 1955–58, TMI), which is now also available, in English, on the Web,[2]
as well as the Aarne-Thompson-Uther classification of folk tales (ATU, see Uther,
2004), of which excerpts in various languages are available in Wikipedia.[3] The

Thierry Declerck
DFKI GmbH, Stuhlsatzenhausweg 3, D-66123 Germany & ICLTT, Sonnenfelsgasse 19/8, 1010
Wien, Austria. e-mail: declerck@dfki.de

Karlheinz Mörth · Gerhard Budin
ICLTT, Sonnenfelsgasse 19/8, 1010 Wien, Austria. e-mail: {Karlheinz.moerth,gerhard.
budin}@univie.ac.at

Piroska Lendvai · Tamás Váradi
HASRIL, 1068 Budapest, Benczúr u. 33, Hungary. e-mail: {piroska,varadi}@nytud.hu

[1] See http://linkeddata.org

[2] See http://www.ruthenia.ru/folklore/thompson/index.htm

C. Chiarcos et al. (eds.), *Linked Data in Linguistics*,
DOI 10.1007/978-3-642-28249-2_11, © Springer-Verlag Berlin Heidelberg 2012

longer term goal of our work is to obtain a LOD-compliant representation not only for the (hierarchy of) classes used in those resources, but also for the language data associated with the classes, and so to establish semantic links between the language data and the classes, across languages and classification systems. A pre-requisite for achieving this stage is the linguistic processing of the content of the labels, and the representation of this analysis in terms of standards compliant with the LOD.

The publication of the resulting domain knowledge and enriched language resources in the LOD can support the automated analysis of literary works by advanced knowledge-driven Natural Language Processing, putting at the disposal of scholars a large set of linguistically and semantically annotated multilingual text segments. Not only researchers in the Humanities will benefit from such a transfer of Humanities resources onto the LOD, but also the general public can be offered extended search possibilities. Folk-literature in general is a very popular topic: as the online presence of the French National Library states, the fables of Jean de la Fontaine are the most consulted literary works in their catalogues;[4] the maintainers of the online Dutch database of folk tales at the Meertens Institute[5] likewise report a high number of visitors.

Such institutions are highly interested in getting machine readable and processable versions of the type of classification systems we mentioned above in order to semi-automatically improve their own indexing, across versions of literary works in different languages and cultures. This would for example allow to detect the differences in the various "national" versions of classical tales over time, and how those are leading to different types of interpretation of the story.

The goal of our work is to provide a resource type that is potentially complementary to the ones already published by national libraries in the LOD.[6] By these, mainly the bibliographical metadata, including the structured part of classification systems, have been ported to the LOD, and we are interested in making all the linguistically and semantically analysed language data included in those classification

---

[3] See http://en.wikipedia.org/wiki/Aarne-Thompson_classification_ system for the English version, http://de.wikipedia.org/wiki/Aarne-Thompson-Index for the German one, and for the French version http://fr.wikipedia.org/ wiki/Classification_Aarne-Thompson. We note that the online ATU data do not reflect the original catalogue, as this was the case for TMI. Therefore it would be highly appreciated to dispose over an electronic version of ATU.

[4] See http://data.bnf.fr/

[5] See http://www.verhalenbank.nl/

[6] See for example the press release of the Hungarian National Library: http://lists.w3. org/Archives/Public/public-lod/2010Apr/0155.html, or for the German National Library (DNB): http://www.eco4r.org/workshop2010/eco4r_workshop2010_ mirjam_kessler.pdf

systems LOD compliant.[7] At the end we will generate a kind of matrix of classification items and the natural language expressions that are typically associated to the classes. This will improve the automatic classification of literary works, on the basis of the automatically combined linguistic and semantic analysis of the texts, detecting variants of the textual expressions used in the labels of the classes.

In doing so we hope to contribute to the "Global Cultural Graph" that is emerging in the LOD.

2 Language Data in the Linked Open Data Framework

Our vision of Linked Language Data (LLD) is to aim at making a better organisation and use of the language data that is (to be) incorporated in the linked open data (LOD) framework. While the LOD initiative was primarily conceived as a way of interconnecting structured knowledge resources in a standardized way using specific links between dereferenceable URIs, the wealth of language data, existing as the content of labels or comments, intuitively associated with those knowledge resources has not been exploited till now, although the LOD framework offers a configuration that can be used and extended for knowledge-driven organization of language data in a set of NLP applications. Additionally, the corresponding linguistic information (e.g. lemma, part-of-speech, morphology, constituency, argument structure, etc.) for those language data is missing, which poses a serious obstacle to direct re-usability in NLP applications.

Adding linguistic information to the language data in LOD will lead to the generation of a very large and dynamically increasing set of enriched language data being structured with both (domain) semantics and linguistic information. This integrated resource can positively impact a range of applications that build on combinations of world and linguistic knowledge, e.g. in Cross-Lingual Information Extraction and Summarization, Localization, Machine Translation, etc.: The knowledge represented in the LOD data sets will become associated with the language data according to linguistic parameters, and thus ready to be directly included in Natural Language Processing modules.

The focus of LLD is thus on bridging the gap between raw language data and knowledge descriptions, by solving representation issues and inclusion of multilingual knowledge-based terminological resources in the LOD. We started to experiment on those ideas with resources in the Humanities, like *Thompson's Motif Index* (TMI) mentioned above, that are not yet in the LOD, but which will be directly published in this framework, including the combination of linguistic and semantic information, as a result of our work.

---

[7] We note for example that the transformation of the index-terms of the DNB onto SKOS is not proposing any additional linguistic categorisation of the terms. See http://www.kim-forum. org/material/pdf/BPG_Repraesentation_von_KOS_im_Semantic_Web.pdf

3 Two Classification Systems for Folk-Literature: TMI and ATU

As mentioned in the introduction section, we are for now dealing with two existing extended catalogues that hold conceptual schemes for classifying narrative elements in folktales, ballads, myths, and related genres: the *Thompson's Motif Index* (TMI, see Thompson, 1955–58) and *The Types of International Folktales* (ATU, see Uther, 2004). ATU and TMI come along with extensive terminologies, and our idea is that these can be linguistically processed, enriched, linked, and represented for language technology mechanisms to identify the semantic classes that can be associated with a text. Pursuing this line of research, we soon discovered that digital catalogues require important terminological and semantic pre-processing in order to be successfully used as a knowledge base to be matched with textual data. Furthermore, catalogues have to be made interoperable with each other, and we need to map/transform them into a semantically harmonized representation, using standards such as XML-TEI or RDF.

We focus therefore on porting TMI and ATU into adequate semantic resources, under consideration of linguistic and terminological aspects. A strong wish in the Digital Humanities community lies in linking ATU and TMI, and we think that this can be carried out only by providing for multilingual and semantic extensions of those resources.

3.1 Work on the Thompson Motif-Index

The Thompson Motif-Index is an hierarchical structure of motifs descriptions consisting in a label associated with an alphanumeric class index, so for example: "A21.1 :: Woman who fell from the sky". Higher in the hierarchy are:"A0 :: Creator", "A20 :: Origin of the Creator" and "A21 :: Creator from above".

We propose both a lexical and a syntactic analysis of all those labels, and for our example A21.1. we get[8]:

```
woman,N+Nb=s+Distribution=Hum
who,PRO+Distribution=RelQ
fell,fall,V+Tense=PT+Pers=3+Nb=s
from,PREP
the,DET
sky,N+Nb=s
```

Lexical Analysis of the label: Woman who fell from the sky.

```
<NP>
    <NP>
        <HEAD><REFOF XREF="396.2">Woman</HEAD>
    </NP>
```

---

[8] This analysis is the result of a specialized grammar we wrote using the NooJ platform, see www.nooj4nlp.net/.

```
<SENT>
    <RELCLAUSE>
        <SUBJ><XREF>who</XREF></SUBJ>
        <PRED>fell</PRED>
        <PP><PPOBJ>from
            <NP>
                <SPEC>the</SPEC>
                <HEAD>sky</HEAD>
            </NP>
        </PPOBJ></PP>
    </RELCLAUSE>
</SENT>
</NP>
```

Analysis of the label: Woman who fell from the sky.

In our analysis, we also make use of the Conceptual, Terminological and Linguistic Objects in Ontologies (CTL) approach, described in Declerck and Lendvai (2010), and which consists in keeping track of the relations between annotated labels of knowledge systems and the classes or properties they are associated with. For the the linguistic level, we provide syntactic information on both the constituency (phrasal grouping of word forms) and the dependency relations between (groups of) word forms. For example, *woman* and *the sky* are marked as a constituent of type NP. At the dependency level, *woman* is marked as the "Subj(ect)" of the "Pred(icate)" *fell*. The detection of this Subj-Pred relation is possible only on the basis of an earlier round of computing the co-reference relation between *woman* and *who*, which we marked with the XML element "XREF". We also mark the dependency relation between the word forms within a phrasal constituent. This allow us to create a "lexicalized ontology" stating for example that a *woman* can be a *creator*, that the *sky* can be related to the *above*, on the base of the lexical and syntactic realisations of concepts included in the hierarchy of TMI.

But we note that the NooJ encoding of the CTL mechanisms is lacking inter-operability with other resources. Therefore actual work is dedicated in linking the NooJ annotation to the ISO data categories,[9] and to use the *lemon* model developed in the Monnet project.[10] for representing the different kinds of data concerned – conceptual, terminological and lexical , and with this to test if we can publish the result of our work directly as a LOD data set. Further we are aiming at linking results of the CTL-*lemon* driven semantic annotation of folktale text to actual LOD data sets published by libraries, relating thus semantic annotation of literary text to bibliographical metadata.

---

[9] See http://www.isocat.org/, also Windhouwer and Wright (this vol.).

[10] See http://www.monnet-project.eu/lemon, also McCrae et al. (this vol.).

3.2 Towards a Multilingual and Semantic Extension of TMI

Analysing the on-line version of TMI, we discovered very soon that it would be beneficial for further automatic processing to turn the basic classification of TMI into a real taxonomy. The actual alphanumeric organization of TMI, which simulates the class hierarchy of motifs does not allow to properly express the hierarchy and inheritance properties of motifs. Furthermore, it is not made explicit in TMI which elements introduce pure classification information ("A0-A99. Creator", "A20. Origin of the creator"), which ones are abstractions over concrete motifs ("A21. /Creator from above./"), and which ones are in fact the actual motifs, i.e. their possible concrete realisation in text ("A21.1. /Woman who fell from the sky./–Daughter of the sky-chief falls from the sky, is caught by birds, and lowered to the surface of the water. She becomes the creator.–*Iroquois: Thompson Tales n.27.–Cf. Finnish: Kalevala rune 1."). We therefore wrote a script that transforms the digital version of TMI onto an XML representation that marks this kind of information explicitly by using designated tags:

```
<CLASS ID="0" SPAN="0-99" LABEL="Creator">
<CLASS ID="20" LABEL="Origin of the creator" SubClassOf="0">
<CLASS ID="21" MOTIF="Creator from above" TYPE="Abstract"
       SubClassOf ="0">
<CLASS ID="21.1" MOTIF="Woman who fell from the sky"
       TYPE="Ref" PartOf="21">
```

As a next step, ongoing work is dedicated to upgrading the XML representation to RDF and OWL, so that we have the adequate means for differentiating between hierarchical realisations and real properties associated with classes, and the possibility to compute the transitive closure of the subclass hierarchy. In parallel, we targeted the extension of motifs listed in TMI in English into a multilingual version. This is carried out by accessing the multilingual Wiktionary[11] lexicon, and suggesting multilingual equivalents to the motifs formulated in in English. Using the lexvo[12] service (available in the LOD framework), one is getting access to Wiktionary and other sources. Lexvo suggests for example for the English word "creator" more than 20 translations. Note that this approach is limited to word-based translations. Our previous study in Mörth et al. (2011) analyses some shortcomings of the use of Wiktionary in its actual state, and proposes a conversion of the Wiktionary lexicon into a TEI compliant format.[13]

---

[11] See http://en.wiktionary.org/wiki/Wiktionary:Main_Page

[12] See http://www.lexvo.org

[13] See http://corpus3.aac.ac.at/showcase/index.php/wiktionary001 for a demo of the transformation of the German wiktionary.

3.3 Towards a Multilingual Combination of ATU and TMI

The complete TMI is available on the web. It is available only in English, as we already mentioned. ATU presents a different situation: only segments of ATU are available on-line in Wikipedia, but in different languages. Looking at the English, French and German Wikipedia pages some discrepancies in the presentation become evident, as illustrated below:

```
(EN) Rapunzel 310 (Italian, Italian, Greek, Italian)
(DE) AaTh 310 Jungfrau im Turm KHM 12 Rapunzel
(FR) AT 310: La Fille dans la tour (The Maiden in the Tower)
     : version allemande
```

The English (EN) version links the German tale *Rapunzel* to four tale versions, in different languages. The original German tale is reached from the English Wikipedia page if the reader clicks on the name "Rapunzel". The German (DE) version links additionally to a German classification system (KHM = Kinder- und Hausmärchen – Children's and Household Tales–, used for the classification of Grimm's Fairy Tales). Interesting enough is the fact that only the French (FR) Wikipedia page introduces the English name for the tale. None of the three Wikipedia pages is making use of the same abbreviation of the ATU index ("'310'", "'AaTh'", or "'AT310'"). Therefore we suggest a restructuration of the information availble in the three pages, merging it in one XML format:

```
<ATU ID="310">
  <LABEL lang="EN">Rapunzel</LABEL>
  <LABEL lang="DE">Jungfrau im Turm</LABEL>
  <LABEL lang="FR">La Fille dans la tour</LABEL>
  <ALT lang="DE">Rapunzel</ALT>
  <ALT lang="EN">The Maiden in the Tower</ALT>
</ATU>
```

This representation will also be ported to RDF, using the SKOS standard for encoding the preferred and alternative forms. On the basis of a small fragment of correctly aligned Wikipedia pages, a representative multilingual terminology of ATU terms can be aggregated, and this terminology can be re-used for supporting the translation of motifs used in TMI, overcoming the shortcomings of the word-based translation approach we discuss in Sect. 3.2. We started to utilize terminology alignment techniques used in Machine Translation (see for example Federmann et al., 2011), adapting them to the short terms that are employed in the catalogues we are focusing on.

As mentioned already, a promising approach of this project is the design of the RDF-based *lemon* representation model for lexicon entries used in ontologies. We are starting to investigate in which way the Monnet project and its *lemon* model can help in translating the English labels of TMI/ATU on the one side and to publish the combination of linguistic information included in the labels of TMI/ATU and their corresponding classes in a LOD compliant way, see also McCrae et al. (this vol.).

4 Conclusion

We have presented actual work dealing with classification systems in the field of eHumanities, with the goal of "upgrading" those resources into interoperable multilingual systems, taking also into consideration linguistic and terminological issues. Next step of our work will be in proposing a LOD format for the resulting combined linguistically and semantically analysed classification data.

Acknowledgements The work described in this paper is done in the context of a collaboration between the Academy of Sciences of Budapest and Vienna. The contribution by Thierry Declerck is partly supported by the R&D project "Monnet", which is co-funded by the European Union under Grant No. 248458.

References

Declerck T, Lendvai P (2010) Towards a standardized linguistic annotation of the textual content of labels in knowledge representation systems. In: Proceedings of the 7th International Conference on Language Resources and Evaluation (LREC'10), Valetta, Malta

Federmann C, Hunsicker S, Wolf P, Bernardi U (2011) From statistical term extraction to hybrid machine translation. In: Proceedings of the 15th Annual Conference of the European Association for Machine Translation

McCrae J, Montiel-Ponsoda E, Cimiano P (this vol.) Integrating WordNet and Wiktionary with *lemon*. pp 25–34

Mörth K, Declerck T, Lendvai P, Várdi T (2011) Accessing multilingual data on the web for the semantic annotation of cultural heritage texts. In: Proceedings of the 2nd International Workshop on the Multilingual Semantic Web

Thompson S (1955–58) Motif-index of folk-literature: A classification of narrative elements in folktales, ballads, myths, fables, medieval romances, exempla, fabliaux, jest-books, and local legends. Indiana University Press, Bloomington

Uther HJ (2004) The Types of International Folktales: A Classification and Bibliography. Based on the system of Antti Aarne and Stith Thompson. Suomalainen Tiedeakatemia, Helsinki

Windhouwer M, Wright SE (this vol.) Linking to linguistic data categories in ISOcat. pp 99–107

OntoLingAnnot's Ontologies: Facilitating Interoperable Linguistic Annotations (Up to the Pragmatic Level)

Antonio Pareja-Lora

Abstract This paper presents the OntoLingAnnot annotation framework, already developed for the annotation of morphological, syntactic, semantic and discourse phenomena, and its extension to cover the annotation of pragmatic phenomena. This extension was considered the ideal test bed for the interoperability of the linguistic annotations performed by means of the platform, since (i) pragmatics itself deals with a real mix of different linguistic topics, such as speech acts, pragmatic coherence relations, deixis, presuppositions and implicatures; and (ii) it clearly interacts with the rest of levels, since (potentially) every linguistic unit at any level can have a pragmatic projection. In particular, it introduces the different pragmatic units that can be used to annotate texts and dialogues using the framework. These pragmatic units are included in the set of ontologies associated to OntoLingAnnot, whose design requirements and development process are also described here. Besides, this paper shows as well the main principles and properties of the OntoLingAnnot annotation framework that help its different annotations interoperate.

1 Introduction

Linguistics and linguistic annotation are very wide fields and, due to this, they have been traditionally partitioned somehow for their study and/or research. Therefore, the most usual criterion to partition them is based on the concept of level, which divides Linguistics into, for example, morphology, syntax, semantics, discourse and/or pragmatics. This partition of Linguistics and its applications has given rise to several good separate models of its different levels, which, nonetheless (and unfortunately), cannot interoperate and do not benefit from the advances of the others in most of the cases. This is due to the fact that this partition has also led to a some-

Antonio Pareja-Lora
ILSA-UCM / ATLAS-UNED, Facultad de Informática (UCM) – Profesor José García Santesmases s/n – 28040-Madrid. e-mail: apareja@sip.ucm.es

C. Chiarcos et al. (eds.), *Linked Data in Linguistics*,
DOI 10.1007/978-3-642-28249-2_12, © Springer-Verlag Berlin Heidelberg 2012

what poor communication between the resulting subareas and a lack of a global perspective of Linguistics and, in particular, of linguistic annotation.

Much on the contrary, the OntoLingAnnot framework has been developed following a comprehensive and complementary approach, which considers all these levels of annotation together, not separately. As commented in Pareja-Lora and Aguado de Cea (2010), this comprehensive approach allowed comparing these different levels and "finding the differences and similarities between them, so as to bear a general and uniform (level-independent) annotation framework across levels. In this comparison process, some regularities and uniformities across levels were found, which help structure and formalize all of them". These regularities and uniformities also help define more interoperable linguistic annotation schemes, since they rely on the common and level-independent properties of linguistic categories, and not so much on their common and level-dependent properties.[1]

This paper presents OntoLingAnnot and its extension in order to cover the annotation of pragmatic phenomena. In this extension, pragmatics was regarded as the ideal test bed for the interoperability of linguistic annotations. On the one hand, pragmatics itself deals with a real mix of different linguistic topics, such as (i) speech acts (Searle, 1979), (ii) pragmatic coherence relations (Asher and Lascarides, 2003); or (iii) deixis, presuppositions and implicatures (Levinson, 1983). These different pragmatic topics have been tackled traditionally following several fragmentary and/or partial approaches. Thus, developing an overall annotation scheme for pragmatics involved the interoperation of several separate (types of) linguistic annotations. On the other hand, the clear interaction of pragmatics with the rest of levels already included in OntoLingAnnot[2] (since, potentially, every linguistic unit at any level can have a pragmatic projection) entailed making the annotations of the remaining levels interoperate at least with the pragmatic ones.

Accordingly, this paper shows the main principles and properties of the OntoLingAnnot annotation framework that help its different annotations interoperate, together with an important part of the formalization of the pragmatic (annotation) level integrated into this framework. In particular, it introduces the different pragmatic units that can be used to annotate texts and dialogues using the framework. These pragmatic units are included in the set of ontologies (Gruber, 1993; Borst, 1997) associated to OntoLingAnnot, whose design requirements and development process are also described here.

This paper is organized as follows: Section 2 states the background and the main assumptions underlying the OntoLingAnnot annotation framework, distributed between Sect. 2.1, which summarizes OntoLingAnnot's basic principles and components, and Sect. 2.2, which discusses its improvements and own contributions. Then,

---

[1] Our efforts are related to other contributions in this volume, in particular Windhouwer and Wright (this vol.) and Chiarcos (this vol.). Whereas these, however, take a bottom-up perspective on linguistic annotations and register categories from existing annotation schemes (Windhouwer and Wright, this vol.) or generalize over them (Chiarcos, this vol.), OntoLingAnnot takes a top-down perspective in that it provides a formalization of linguistic phenomena which is then linked to annotation schemes.

[2] Morphology, syntax, semantics and discourse.

OntoLingAnnot's pragmatic units are presented in Sect. 3. The results and the conclusions drawn from the development and the evaluation of the framework are commented on in Sect. 4.

2 The OntoLingAnnot Framework

OntoLingAnnot (Pareja-Lora and Aguado de Cea, 2010; Pareja-Lora, in press) is a new annotation framework that applies some of the principles underlying the Onto-Tag annotation model of Aguado de Cea et al. (2002, 2004) and also reuses some of its components. Hence, it is necessary to first introduce the OntoLingAnnot principles and components coming from OntoTag.

2.1 OntoLingAnnot's Principles and Components Coming from OntoTag

OntoTag basically consists of both (1) an annotation architecture and (2) an annotation scheme. On the one hand, OntoTag's annotation architecture allows for the collaboration and interoperation of several tools on the annotation of web pages. On the other hand, OntoTag's annotation scheme aims at an interoperable and joint annotation of morphosyntactic, syntactic and semantic[3] phenomena.

In particular, the principles and components of OntoTag that OntoLingAnnot applies and reuses come from its annotation scheme. They can be enumerated as follows:

1. A clear differentiation between the linguistic data categories (LDCs) used in the annotations and the format (or the way) in which these annotations are expressed. This differentiation contributes largely to enhancing the interoperability of linguistic annotations and is completely in line with the ISO standards being developed for linguistic annotation. LDCs are the object of a standard coming from the ISO/TC37/SC3, namely ISO/TC37/SC3 – Terminology and other language and content resources (2008), cf. Windhouwer and Wright (this vol.), linguistic annotation schemes are the objects of several other standards coming from the ISO/TC37/SC4, such as (1) ISO/TC37/SC 4 – Language resource management (2008) for morpho-syntactic annotation; (2) ISO/TC37/SC 4 – Language resource management (2009c) for syntactic annotation; (3) ISO/TC37/SC 4 – Language resource management (2009b), and ISO/TC37/SC 4 – Language resource management (2010d) for semantic annotation; (4) ISO/TC37/SC 4 – Language resource management (2010b) for discourse annotation; or (5)

---

[3] Restricted to senses and named entities.

ISO/TC37/SC 4 – Language resource management (2010c) for discourse and pragmatic annotation.[4]

2. The formalization of LDCs as ontological terms. Since OntoTag's ontologies are implemented in OWL, this formalization enables identifying and referring to each LDC by means of its own Uniform Resource Identifier (URI), which is one of the requirements included in another ISO/TC37/SC4 standard (ISO/TC37/SC 4 – Language resource management, 2010a).

3. The basic classification of each LDC as a `Linguistic Unit`, or as a component of a linguistic feature, that is, as a `Linguistic Attribute` or as a `Linguistic Value`, in order to avoid redundancy and facilitate modularization.

4. The distribution of LDCs among three main ontologies (Aguado de Cea et al., 2004), originated by the classification described in the previous item: a Linguistic Unit Ontology (LUO), a Linguistic Attribute Ontology (LAO), and a Linguistic Value Ontology (LVO). These three ontologies were linked together by the Integration Ontology (IO), and they four, altogether, are the main components of OntoTag reused in OntoLingAnnot.

5. Texts units are annotated by means of triples `<LinguisticUnit, LinguisticAttribute, LinguisticValue>`, regardless of the type of annotation performed (morpho-syntactic, syntactic or semantic, in the case of Onto-Tag).

6. Text unit annotations, that is, the `<LinguisticSubject, LinguisticAttribute, LinguisticValue>` triples, are implemented by means of RDF triples `<Subject, Predicate, Object>`, in which the corresponding linguistic units (i.e., subjects), attributes (i.e., predicates), and values (i.e., objects), are conveniently formalized as classes or instances of one or more ontologies. This fulfils also one of the main requirements of the standard ISO/TC37/SC 4 – Language resource management (2009a), which is being developed within ISO/TC37/SC4 as well.

Nevertheless, these principles and components of OntoTag proved necessary but insufficient and/or too narrow for OntoLingAnnot from the beginning. The next subsection details how they were amended and extended in order to be generalized.

2.2 OntoLingAnnot's Improvements and Own Contributions

OntoLingAnnot's scope is wider and a bit more ambitious than OntoTag's. It is wider, since OntoLingAnnot tries not only to cover syntax and some phenomena of semantics or in the interface between syntax and morphology (i.e., morpho-syntax);

---

[4] Most surprisingly, ISO/TC37/SC4 has decided to develop these two discourse and pragmatics-related annotation standards in the ISO macro-project for the standards of other indisputable forms of semantic annotations, instead of creating a new project for them, as with ISO/TC37/SC 4 – Language resource management (2008) and ISO/TC37/SC 4 – Language resource management (2009c), for example.

instead, it seeks to cover these three levels completely, as well as discourse and pragmatics. It is more ambitious, since it aims at (a) achieving a real and full inter-operability of its annotations and (b) being flexible and customizable, that is, both (b.1) extendable (to Phonology or Prosody, for example) and (b.2) scalable.[5] In other words, it should be possible to derive any particular annotation scheme from On-toLingAnnot by simply adding to the model and/or selecting from it the particular set of LDCs that refer to the phenomena being annotated. This is why the principles and components of OntoTag inherited by OntoLingAnnot proved insufficient (or too narrow) from the beginning and, hence, some of them had to be adapted as follows.

Firstly, the basic classification of LDCs in OntoTag neglected an important group of linguistic categories, that is, linguistic relationships. This is due to the fact that the few linguistic relationships contemplated in OntoTag (e.g., Syntactic Function or Syntactic Dependency) could be managed as linguistic at-tributes. However, when the set of LDCs of OntoTag was extended to cover mor-phology, discourse and pragmatics, it was clear that linguistic relationships deserve to be treated separately from linguistic attributes. It was also clear that (i) some of them share several properties (even across levels: for instance, the relative rank of the elements they interrelate), which could help distinguish several particular classes of linguistic relationships; (ii) most of them could be arranged into a taxonomy of linguistic relationships (e.g., dependencies, functions and coherence relations (see Hovy and Maier, 1995); and (iii) some of them possess their own attributes, which help characterize and subclassify them.[6] These three arguments made of linguis-tic relationships potential concepts of an ontology. This is how a fifth ontology of linguistic relationships (the Linguistic Relationship Ontology, LRO) came into be-ing and joined immediately the ones present in OntoTag. Linguistic attributes were searched afterwards in order to tell real linguistic attributes from linguistic rela-tionships and move the latter to the LRO. Then, the LRO was swelled with the relationship LDCs that were found when (re-)formalizing the five levels mentioned above.

Secondly, as soon as linguistic relationships were put into play in OntoLingAn-not, it was necessary to reconsider the types of linguistic triples defined in OntoTag. Thus, a new type of triple, <LinguisticUnit, LinguisticRelation, LinguisticUnit>, was added to the <LinguisticUnit, Linguis-ticAttribute, LinguisticValue> triples inherited by the framework.

Thirdly, the linguistic triples discussed in the previous paragraphs, when im-plemented, originated two types of RDF <Subject, Predicate, Object>

---

[5] Although no exhaustiveness can be claimed, there is a huge amount of linguistic terms (around 2000) already formalized in OntoLingAnnot's ontologies. Therefore, for certain linguistic phe-nomena and for some reasons (for instance, in order to guarantee a satisfactory inter-annotator agreement) it might be useful and recommended to reduce the set of linguistic categories chosen for their annotation. This can be easily done by ignoring some of the more fine-grained (though pertinent) terms in the ontologies, and choosing only the more coarse-grained.

[6] For instance, according to Hovy and Maier (1995), coherence relations are characterized by being established at the discourse or at the pragmatic level, which helps characterize and subclassify co-herence relations into two different classes: discourse coherence relations and pragmatic coherence relations.

triples, namely (i) those in which the Predicate is a Linguistic Attribute and the Object is a Linguistic Value; and (ii) those in which the
Predicate is a Linguistic Relationship and the Object is a Linguistic Unit.[7] Since all of them were already formalized as classes or instances
of OntoLingAnnot's ontologies, this did not diminish the compliance of the model
with the standard ISO/TC37/SC 4 – Language resource management (2009a).

2.2.1 OntoLingAnnot's Annotation Processes and/or Layers

Finally, up to this point, LDCs had been classified according to two different criteria
or axes, that is, (1) the level to which they belong (morphology, syntax, semantics,
discourse and/or pragmatics) and (2) the type of category they constitute (a Linguistic Unit, a Linguistic Attribute, a Linguistic Value or a
Linguistic Relationship). Then, they were also classified according to a
new criterion, i.e., the annotation process in which they are handled. In effect, some
annotation processes were identified when studying the regularities and uniformities that share the different annotation models developed thus far (independently of
their level). Following the terminology in Leech et al. (1996), each of these annotation processes was referred to as an Annotation Layer, and each of the LDCs
was linked to the Annotation Layer in which it is used for annotation. The
resulting layers can be described as follows.

The first layer is the Segmentation Layer. In this layer, the linguistic units
that are to be annotated are firstly identified and delimited, segmenting thus the text
into its constituent units (according to the level considered).

The second layer is the Paradigmatic Labelling Layer, in which the
units segmented in the previous layer can be further characterized by sub-classifying
them and/or accompanying them with the particular features (i.e., the pairs <At-
tribute, Value>) of the level in question that they present in the text.

The third layer is the Syntagmatic Relation Identification Layer.
To improve the annotation of the text, the linguistic relations holding between the
linguistic units at the level considered can be identified as well. This layer is in
charge of this type of annotation.

The fourth layer, the Syntagmatic Relation Labelling Layer, is
responsible for further (and optionally) refining the annotation of these relations,
as with units, sub-classifying and/or characterizing them by means of their corresponding features in the text.

The fifth layer is called the Resulting Unit Layer. A full annotation of
a text at a given level includes, apart from the annotation of the layers mentioned
above, an optional and complementary annotation of the higher-rank units that result
from the composition or aggregation of other units, by means of one or more relations of that level (already identified and annotated). This is performed, at each level,
within its particular Resulting Unit Layer. Since linguistic levels cannot be

---

[7] The Subject is a Linguistic Unit in both cases.

considered disjoint, the units that constitute the interface between two or more levels must be detailed in this layer too. As shown in the development of OntoLingAnnot, in most cases, the annotation of the units on the interface between two levels is a critical aspect as for the interoperability of linguistic annotations.

This classification of LDCs according to their layer helps scale and customize OntoLingAnnot to the needs of each particular linguistic annotation project. Obviously, depending on how deep the annotations need to be, they will include more or less layers of annotation, since they are fairly independent. Once decided which layers are to be annotated, the LDCs (i.e., the terms) available in OntoLingAnnot's ontologies for the annotation of these layers can be automatically selected and extracted. This frees the users of the platform from having to browse the whole set of OntoLingAnnot's LDCs in search of their own linguistic data category selection for a certain type of linguistic annotation.

All these principles, components and improvements constitute the backbone of the OntoLingAnnot annotation framework and its main contributions. The following section summarizes the pragmatic units that it includes. They represent the most relevant LDCs that can be used for the pragmatic annotation of texts according to this framework. The rest of classes and instances included in the pragmatic modules of OntoLingAnnot's ontologies (namely its pragmatic units, pragmatic attributes, pragmatic values and pragmatic relationships) are discussed in Pareja-Lora (in press). They are not included here for the sake of space.

3 The Pragmatic Units of OntoLingAnnot

The main (i.e., top-level) classes of the Linguistic Unit Ontology (LUO) that formalize the pragmatic units contemplated in OntoLingAnnot are Macroproposition, Pragmateme and Pragmatic Functional Unit.

A Macroproposition is both a Pragmatic Unit and a complex Discourse Unit[8] that serves as a unitary construction block at the Pragmatic Level. A Speech Act, for example, is a type of Macroproposition, as well as a Trope (e.g., a Metaphor). Macropropositions can be regarded as the linguistic units that result from the aggregation of some interrelated propositions from the Discourse Level (Dijk, editor). The Apology ('Excuse me'), the Query ('can you tell me where the nearest police station is, please?') and the Begging Act ('please') in Example 1 are instances of this type of units.

> Person A: *Excuse me, can you tell me where the nearest police station is, please?*
> Person B: *Go down the street and turn left at the traffic lights. I think it's on the right.*

Example 1: An excerpt of a short dialog

---

[8] That is, macropropositions stand on the discourse-pragmatics interface.

Macropropositions can be related to each other by means of pragmatic relations in order to build pragmatemes. The unit `Pragmateme`, hence, represents in OntoLingAnnot the main resulting unit of a text pragmatic analysis. The role of this kind of `Pragmatic Unit` in pragmatic annotation can be better understood in the light of the units into which it can be sub-classified, such as `Macroproposition Aggregation Pragmateme (MAP)`, `Pragmatic Transposition Unit (PTU)`, `Emphasis-Related Unit (ERU)`, `Saying` or `Set Phrase`. Example 1, as a whole, constitutes a particular type of `Pragmateme`, i.e., a `Macroproposition Aggregation Pragmateme`, which consists of a `Query` and its corresponding `Answer`, linked at the `Pragmatic Level` by a type of `Adjacency Pair Relation`.

Finally, a `Pragmatic Functional Unit (PFU)` signals a `Pragmatic Coherence Relation` (see Pareja-Lora, in press). Thus, a `PFU` is a linguistic unit that makes explicit a pragmatic relation that holds between two (adjacent) pragmatemes in text or in dialogue. This unit extends the concept of `Discourse Functional Unit (DFU`, see Romera, 2004) to the `Pragmatic Level`. For this reason, PFUs are to the `Pragmatic Level` and to pragmatic coherence relations as DFUs are to the `Discourse Level` and to discourse coherence relations. The change of `Turn` in Example 1 is an instance of an `Answer PFU`.

In total, OntoLingAnnot contains 192 pragmatic units (in the Linguistic Unit Ontology, LUO). As for the rest of ontological terms concerning pragmatics in this platform, briefly, OntoLingAnnot's ontologies contain

- 26 pragmatic attributes - 10 concepts and 16 instances (in the Linguistic Attribute Ontology, LAO);
- 81 pragmatic values - 27 concepts and 54 instances (in the Linguistic Value Ontology, LVO);
- 86 classes of pragmatic relations (in the Linguistic Relationship Ontology, LRO); and
- 24 pragmatic concepts relating the pragmatic level and its layers (in other OntoLingAnnot's ontologies).

They amount to 409 pragmatic terms: 339 concepts and 70 instances, apart from several other ontological terms (attributes, *SubclassOf*, *PartOf* and ad hoc relations, rules and axioms).

4 Results and Conclusions

This paper has introduced the OntoLingAnnot (linguistic) annotation framework and also its pragmatic units as a way to show its potential for pragmatic annotation and its interoperability with other levels of annotation. As shown in Pareja-Lora (in press), this is the first ontological (and, hence, computable) conceptualization of pragmatics thus far and, hence, it is an important contribution per se to the areas of Ontological Engineering, Pragmatics and Linguistic Annotation. Besides, no other pragmatic model or framework accounts globally and coherently for such a number of pragmatic phenomena and categories as those formalized and included in OntoLingAnnot's ontologies, which is another important contribution to the areas aforementioned.

There remains the issue of the compliance of OntoLingAnnot with the ISO standards developed so far (namely ISO/TC37/SC 4 – Language resource management, 2008, ISO/TC37/SC 4 – Language resource management, 2009c, ISO/TC37/SC 4 – Language resource management, 2010a, and ISO/TC37/SC 4 – Language resource management, 2009a), which was sought and evaluated all throughout its development. The results of this continued evaluation were fairly satisfactory, in particular as far as the respective LDC coverage of each of its levels was concerned. As for the compliance of OntoLing's pragmatic annotations with ISO standards, the only standard dealing (tangentially) with pragmatics, as understood in this framework, is ISO/TC37/SC 4 – Language resource management (2010c). This standard draft contains a section dealing with speech acts (or dialogue acts, as they are termed in this document). When compared, OntoLingAnnot contains not only the categories for dialogue acts mentioned in ISO/TC37/SC 4 – Language resource management (2010c), but also some others collected mainly from the usual terminology of politics, law and religion (for the extension of commissives and declarations) as well as from several dictionaries (for the extension of directives and expressives). In addition, since speech acts are only a particular type of macropropositions and there are many other types of pragmatic units in OntoLingAnnot, at least its terminological coverage clearly exceeds the one of ISO/TC37/SC 4 – Language resource management (2010c).

Taking into account that, as shown previously, this approach can also be considered flexible, scalable, extensible and, thus, highly (re)usable, OntoLingAnnot can be viewed as an alternative reference model for the development of future linguistic and interoperable annotations.

Acknowledgements This research has been partially (i.e., financially) supported by Banco Santander and UCM, within the GR42/10-962022 project. The author also wants to thank, for their support when carrying out this research, the Engineering of Software Languages and Applications (Ingeniería de Lenguajes Software y Aplicaciones, ILSA) research group, from the Universidad Complutense de Madrid (UCM), and the Artificial Intelligence Techniques for Linguistic ApplicationS (ATLAS) research group, from the Universidad Nacional de Educación a Distancia (UNED).

References

Asher N, Lascarides A (2003) Logics of conversation. Cambridge University Press, Cambridge, UK

Borst WN (1997) Construction of engineering ontologies. PhD thesis, University of Twente, Enschede, Netherlands

Aguado de Cea G, Gómez-Pérez A, Álvarez de Mon I, Pareja-Lora A, Plaza-Arteche R (2002) OntoTag: A semantic web page linguistic annotation model. In: Semantic Web Meets Language Resources. AAAI Technical Report WS-02-16, AAAI Press, Menlo Park, California, USA, pp 20–29

Aguado de Cea G, Gómez-Pérez A, Álvarez de Mon I, Pareja-Lora A (2004) Onto-Tag's linguistic ontologies: Improving semantic web annotations for a better language understanding in machines. In: ITCC '04: Proceedings of the International Conference on Information Technology: Coding and Computing (ITCC'04), Volume 2. IEEE Computer Society, Washington, DC, USA, pp 124–128

Chiarcos C (this vol.) Interoperability of corpora and annotations. pp 161–179

Dijk (editor) T (1997) Discourse Studies (2 vols.). Sage, London, UK

Gruber TR (1993) A translation approach to portable ontologies. Knowledge Acquisition 5(2):199–220

Hovy E, Maier E (1995) Parsimonious or Profligate: How Many and Which Discourse Structure Relations? Tech. rep., Information Sciences Institute, University of Southern California, URL http://www.isi.edu/natural-language/people/hovy/papers/93discproc.pdf

ISO/TC37/SC 4 – Language resource management (2008) Morpho-syntactic annotation framework (MAF). International Standard Draft: ISO/DIS 24611, International Organization for Standardization (ISO)

ISO/TC37/SC 4 – Language resource management (2009a) Linguistic annotation framework (LAF). International Standard Draft: ISO/DIS 24612, International Organization for Standardization (ISO)

ISO/TC37/SC 4 – Language resource management (2009b) Semantic annotation framework (SemAF) – Part 1: Time & events. International Standard Draft: ISO/DIS 24617-1, International Organization for Standardization (ISO)

ISO/TC37/SC 4 – Language resource management (2009c) Syntactic annotation framework (SynAF). International Standard Draft: ISO/DIS 24615, International Organization for Standardization (ISO)

ISO/TC37/SC 4 – Language resource management (2010a) Persistent identification and sustainable access (PISA). International Standard, Final Draft: ISO/FDIS 24619, International Organization for Standardization (ISO)

ISO/TC37/SC 4 – Language resource management (2010b) Semantic annotation framework (SemAF) – Discourse structures. New Working Item: ISO/PWI 24617-5, International Organization for Standardization (ISO)

ISO/TC37/SC 4 – Language resource management (2010c) Semantic annotation framework (SemAF) – Part 2: Dialogue acts. International Standard Draft: ISO/DIS 24617-2, International Organization for Standardization (ISO)

ISO/TC37/SC 4 – Language resource management (2010d) Semantic annotation framework (SemAF) – Static spatial information. New Working Item: ISO/PWI 24617-6, International Organization for Standardization (ISO)

ISO/TC37/SC3 – Terminology and other language and content resources (2008) Specification of data categories and management of a Data Category Registry for language resources. International Standard Draft: ISO/DIS 12620.2, International Organization for Standardization (ISO)

Leech G, Barnett R, Kahrel P, Halteren Hv, Langé JM, Montemagni S, Voutilainen A (1996) Recommendations for the Syntactic Annotation of Corpora. European

Project Deliverable: EAGLES Document EAG–TCWG–SASG/1.8, EAGLES Consortium, URL `http://www.ilc.cnr.it/EAGLES96/segsasg1/segsasg1.html`

Levinson SC (1983) Pragmatics. Cambridge University Press, Cambridge, UK, reprinted as Vol. A of *Computers & Typesetting*, 1986

Pareja-Lora A (in press) The pragmatic level of OntoLingAnnot's ontologies and their use in pragmatic annotation for language teaching. In: Bárcena E, Read T, Arús J (eds) Technological innovation in the teaching and processing of LSPS, Springer, Madrid, Spain, pp 547–574, to appear 2012

Pareja-Lora A, Aguado de Cea G (2010) Modelling discourse-related terminology in OntoLingAnnot's ontologies. In: Bhreathnach U, Barra-Cusack F (eds) Presenting terminology and knowledge engineering resources online: Models and challenges (TKE 2010), Dublin City University, Dublin, Ireland, pp 547–574

Romera M (2004) Discourse Functional Units: the Expression of Coherence Relations in Spoken Spanish. LINCOM, Munich, Germany

Searle J (1979) Expression and meaning: Studies in the theory of speech acts. Cambridge University Press, Cambridge, UK, (reprinted 1999)

Windhouwer M, Wright SE (this vol.) Linking to linguistic data categories in ISOcat. pp 99–107

Using Linked Data to Create a Typological Knowledge Base

Steven Moran

Abstract In this paper, I describe the challenges in creating a Resource Description Framework (RDF) knowledge base for undertaking phonological typology. RDF is a model for data interchange that encodes representations of knowledge in a graph data structure by using sets of statements that link resource nodes via predicates that can be logically marked-up (Lassila and Swick, 1999). The model I describe uses Linked Data to combine data from disparate segment inventory databases. Once the data in these legacy databases have been made interoperable at the linguistic and computational levels, I show how additional knowledge about distinctive features is linked to the knowledge base. I call this resource the Phonetics Information Base and Lexicon (PHOIBLE)[1] and it allows users to query segment inventories from a large number of languages at both the segment and distinctive feature levels (Moran, 2012). I then show how the knowledge base is useful for investigating questions of descriptive phonological universals, e.g. "do all languages have coronals?" and "does every phonological system have at least one front vowel or the palatal glide /j/?" (Hyman, 2008).

1 Introduction

Linked Data is a set of recommended practices for publishing RDF-structured data on the Web (Bizer et al., 2007). RDF is a model for data interchange that encodes representations of knowledge in a graph data structure by using sets of statements (aka 'triples' or 'ontological commitments') that link resource nodes via predicates (Lassila and Swick, 1999; Beckett, 2004). For example the triples, graphically illustrated in Fig. 1, represent a collection of statements where a subject (the language

Steven Moran
Ludwig-Maximilians-Universität, München, Schellingstrasse 9, D-80539 München, e-mail:
steve.moran@lmu.de

[1] http://phoible.org

C. Chiarcos et al. (eds.), *Linked Data in Linguistics*,
DOI 10.1007/978-3-642-28249-2_13, © Springer-Verlag Berlin Heidelberg 2012

Sisaala, Western [ssl])[2] is connected via the relationship hasSegment to the segments p, b, kp, etc.

Fig. 1 Snippet of PHOIBLE
RDF knowledge base

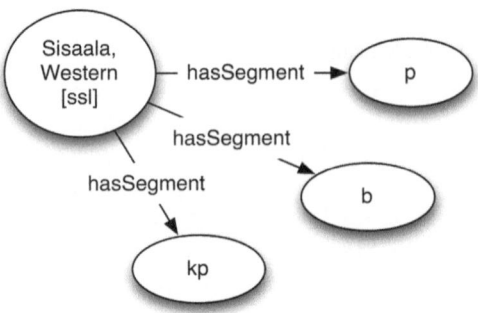

RDF is a data model for defining and specifying resources (here languages and segments) and the relations that hold between them. The product of this model, together with the technologies used to store and access it, is a knowledge base. A knowledge base encapsulates the capabilities of what several integrated technologies allow a user to achieve through the Semantic Web framework.

2 Background

Tim Berners-Lee and colleagues coined the term "Semantic Web" and gave a vision to a "web of data" (Berners-Lee et al., 2001), now commonly referred to as Linked Data. This vision of a web of interlinked Data is motivated by the fact that the Web has evolved mainly as HTML web pages that publish information for human consumption. Their inherent meaning is not interpretable by computers because they lack rich-machine readable metadata. The goal of the Semantic Web vision is to make possible the processing of information published on the Web by computers (Cardoso and Sheth, 2006). In Semantic Web architecture, an application framework stores data in a knowledge base. The Semantic Web is built in layers. A node in the Semantic Web, i.e. a concept, individual or class, is a Uniform Resource Identifier (URI). Triples are built with URIs that define the subject, predicate and object of a statement. Each triple describes a fact. The subject and predicate are defined with a URI. The object of the statement can be either a URI or some other definable data type, such as a string literal or an integer. The URI is a key feature in the overall architecture because each provides a unique identifier within a global namespace. Since triples are built with URIs, they can be easily merged from many different sources via common URIs or defining of relationships between URIs via additional

---

[2] ISO 639-3 language codes are given in brackets [ ].

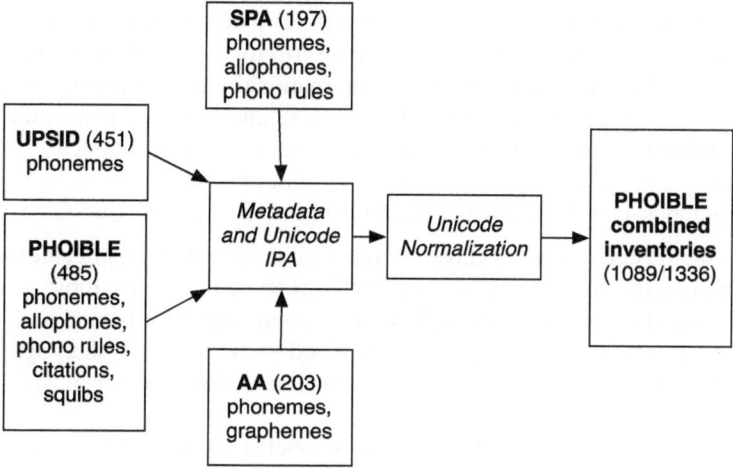

Fig. 2 Resources combined into PHOIBLE

triples. The URI is a key design feature because it provides the mechanism for global naming and connects each resource in the statement to a Web resource.

3 Data and Model

Segment inventory databases have been used since the 1970s to investigate phonological universals and to report on the distribution of sounds in the world's languages. The data used in this paper come from three databases and from my own work extracting segment inventories from the linguistics literature. The first database is the Stanford Phonology Archive (SPA, Crothers et al., 1979). It contains segment inventories (phonemes and allophones) for 197 languages. The second is the UCLA Phonological Segment Inventory Database (UPSID), which contains 451 languages' phoneme inventories (Maddieson, 1984; Maddieson and Precoda, 1990). The third is the 200 language sample (phonemes and graphemes) from *Alphabets des langues africaines* (Hartell, 1993; Chanard, 2006).[3] Additional inventories, extracted from over 400 grammars or phonological descriptions, include phonemes, allophones and phonological rules. Figure 2 illustrates how these resources are combined into the PHOIBLE knowledge base. At the time of writing, there are 1336 segment inventories, which represent 1089 distinct languages.

Combining disparate data sets is challenging. Integrating segment inventories from different resources into one interoperable data set poses three main challenges.

---

[3] Note that the quality of the original and digitized versions of these data were questionable, so Christopher Green and I corrected and/or collected additional resources for these inventories.

The first challenge is how to compare segments from different transcription systems. This issue is addressed by interpreting transcription systems' segments into the International Phonetic Alphabet (IPA, International Phonetic Association, 2005), which I use as an interlingual pivot. However, re-encoding segments into IPA can also be problematic. For example, when more than one diacritic appears to the right of, or below a segment, in which order should they appear (e.g. a creaky voiced syllabic dental nasal / n̰̩ /)? A 'correct' diacritic ordering does not seem to be explicitly stated in the IPA. Therefore, I chose to create an order that all (applicable) segments in all segment inventory databases in PHOIBLE abide by. For example, when there is more than one diacritic to the right of a segment, the order is: unreleased/lateral release/nasal release $\rightarrow$ palatalized $\rightarrow$ labialized $\rightarrow$ velarized $\rightarrow$ pharyngealized $\rightarrow$ aspirated/ejective $\rightarrow$ long, e.g. a labialized aspirated long alveolar plosive is represented as / t^{wh}ː /.[4]

The second challenge involves making segments interoperable at the computational level. For example, a nasalized creaky vowel /ḛ̃/ has diacritics that appear above and below the vowel. Although the order is not visually distinguishable, computationally there are two sequences in which these characters can occur (depending on how they are keyboarded by the linguist).[5] To address this issue, Unicode normalization is needed to decompose each complex segment into an ordered sequence of characters to ensure that equivalent strings have the same binary representation (The Unicode Consortium, 2007).

The third challenge involves associating metadata with each segment inventory in the knowledge base. Each inventory needs to be identified with a language and bibliographic information about the publication from which the inventory was gathered. In my approach, each inventory is identified by an ISO 639-3 language code, which allows inventories representing the same language to be compared systematically, even if the language names used in those resources differ (e.g. *German* vs *Deutsch*). Bibliographic data is associated with the Open Language Archives Community (OLAC) metadata set. OLAC expands the set of DCMI metadata categories to include information pertinent to linguistics data to create a standard way to document many types of language resources, by adding metadata elements like subject language and linguistic data type to enhance greater discovery of language resources.

After these challenges have been addressed to make the data interoperable at the linguistic and computational levels, RDF is used to model languages and their segment inventories. The RDF data structure represents a collection of facts about information using URIs.[6] This knowledge base can be queried with SPARQL, an RDF

---

[4] Ordering conventions are stated explicitly in Moran (2012), as are the SPA and UPSID segment-to-IPA mappings.

[5] There are two possibilities in Unicode: U+0065 + U+0330 + U+0303 (LATIN SMALL LETTER E + COMBINING TILDE BELOW + COMBINING TILDE) or U+0065 + U+0303 + U+0330 (LATIN SMALL LETTER A + COMBINING TILDE + COMBINING TILDE BELOW).

[6] The figures in this paper use terms to represent URIs for readability purposes. However, in the published PHOIBLE RDF model, a segment is defined by an explicit URI, for example `http://phoible.org/segment/kp`, which is consistent with the RDF specification.

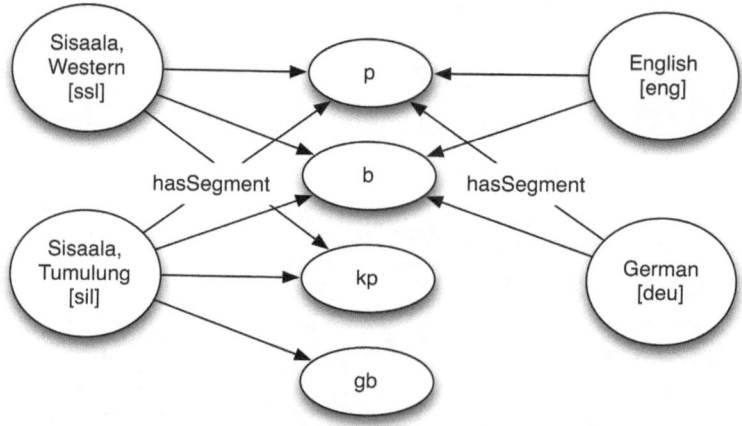

Fig. 3 Expanded snippet of PHOIBLE RDF knowledge base

query language (Prud'Hommeaux and Seaborne, 2006). SPARQL queries consist of triple patterns that match concepts and their relations by binding variables to match graph patterns. For example, a SPARQL query on the knowledge base in Fig. 3 to retrieve the segments of `Sisaala, Western [ssl]` is given in example 2:

(2) `SELECT ?segments`
     `WHERE {ssl hasSegment ?segments}`

The SPARQL query matches any sets of triples that contain `ssl` (for Sisaala, Western) as the subject and `hasSegment` as the predicate. In this snippet the query would return the segments p, b and kp.

An important feature of the RDF graph model, and thus Linked Data, is that the the same knowledge representation language is used in the knowledge base's structure and its encoding of data instances. This is because the knowledge base uses triples to define its structure. Thus, triples can be easily added to by defining new resources (subjects or objects of triples) or predicates. This self-describing structure supports a model of open and shared data. For example, if one wants to add new knowledge about distinctive features to the segment inventory knowledge base, the features can be added to the graph by linking them from each segment via another predicate.[7] For example, I define a URI for the `hasFeature` predicate to link features to segments in Fig. 4. Now if the user wants to query for segments in languages that only contain certain features, they can. Perhaps someone wants to query for all languages that have velar plosive segments. Then they could use the query given in

---

[7] Distinctive feature sets lack typological coverage for the vast variety of segment types that appear in the linguistics literature. In this paper I gloss over the challenges involved in assigning features to segments. See Moran (2012) for details.

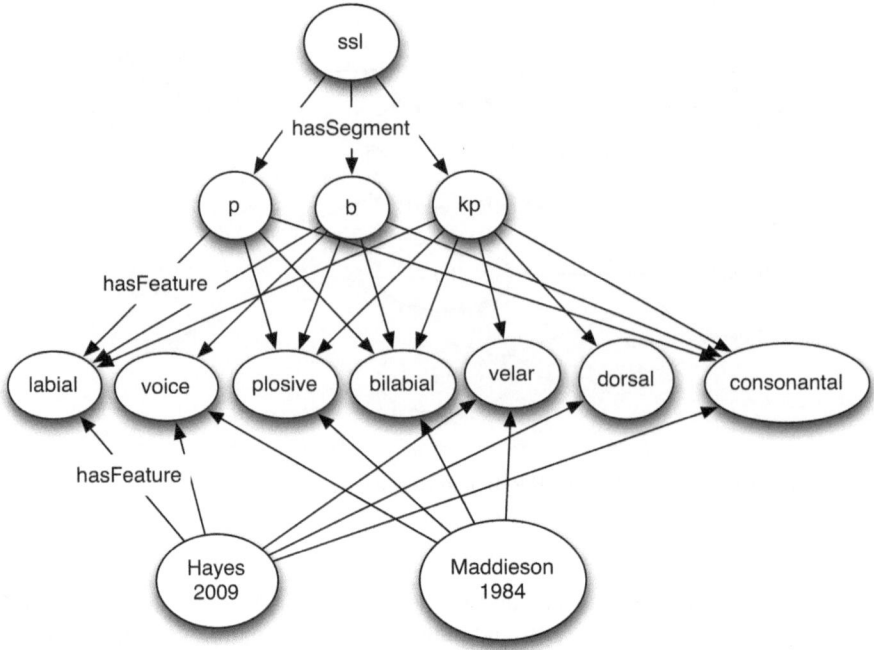

Fig. 4 Adding distinctive features to the knowledge base

example 3. It is now apparent how one might use the PHOIBLE Linked Data graph
to investigate questions of phonological universals.

```
(3)  SELECT ?languages
     WHERE {?languages hasSegment ?segments .
     ?segments hasFeature plosive .
     ?segments hasFeature velar }
```

4 Query

The PHOIBLE Linked Data graph of segments and distinctive features can be used
to investigate descriptive universals of phonological inventories, for example, some
of those stated in Hyman (2008).[8] One area to investigate is vowel systems. Hyman
asks if "every phonological system has at least one unrounded vowel" and reaches

---

[8] See Hyman (2008) for a description of various types of universals. At this time, PHOIBLE
cannot be used to address theory-dependent architectural universals, e.g. statements made within
Optimality Theory, or universals dealing with tendencies above the segment level, e.g. universals
regarding syllable structure.

the conclusion, based on the data from UPSID-451 (Maddieson, 1984; Maddieson and Precoda, 1990),[9] that no language in UPSID-451 has less than two unrounded vowels. This universal can be formulated, as in the SPARQL query in example 4, using features from Hayes (2009).

(4) `SELECT DISTINCT ?languages`
    `WHERE {?languages phoible:hasSegment ?segments .`
    `?segments phoible:hasFeature feature:SYLLABIC .`
    `?segments phoible:notHasFeature feature:CONSONANTAL .`
    `?segments phoible:notHasFeature feature:ROUND }`

This query selects all distinct languages that have segments that have the features [+SYLLABIC], [−CONSONANTAL] and [−ROUND], i.e. unrounded vowels. All 1089 languages are returned. The query can also be modified to return all languages and their segments, shown in example 5.

(5) `SELECT ?languages ?segments`
    `WHERE {?languages phoible:hasSegment ?segments .`
    `?segments phoible:hasFeature feature:SYLLABIC .`
    `?segments phoible:notHasFeature feature:CONSONANTAL .`
    `?segments phoible:notHasFeature feature:ROUND }`

This query returns all language inventories and their unrounded vowels according to our feature set and specification.

Next, by querying the graph for distinct languages that have segments that have the features [+SYLLABIC], [−CONSONANTAL] and [+BACK], we can confirm that Hyman's stated universal, "every phonological system has at least one back vowel", holds in the expanded PHOIBLE dataset.

(6) `SELECT DISTINCT ?languages`
    `WHERE {?languages phoible:hasSegment ?segments .`
    `?segments phoible:hasFeature feature:SYLLABIC .`
    `?segments phoible:notHasFeature feature:CONSONANTAL .`
    `?segments phoible:hasFeature feature:BACK}`

Another possible universal investigated by Hyman is "every phonological system has at least one front vowel or the palatal glide /j/". This can be asked of the PHOIBLE Linked Data graph by using the SPARQL UNION operator to query all languages that have segments of a particular feature make-up ([+SYLLABIC, +ROUND, −CONSONANTAL] or the segment /j/.[10] This universal also holds in the PHOIBLE dataset.

(7) `SELECT DISTINCT ?languages`
    `WHERE {?languages phoible:hasSegment ?segments .`
    `?segments phoible:hasFeature feature:SYLLABIC .`
    `?segments phoible:hasFeature feature:FRONT .`
    `?segments phoible:notHasFeature feature:CONSONANTAL .`
    `UNION {?languages phoible:hasSegment segment:j}}`

---

[9] The data is taken from Henning Reetz's online version, at: http://web.phonetik. uni-frankfurt.de/UPSID.html.

[10] Note I use IPA /j/ instead of /y/.

Another area to investigate descriptive universals in segment inventories is in consonant systems. Hyman posits that "every phonological system has stops" and "every phonological system has coronal phonemes". Example 8 queries for the first universal by selecting all languages that have segments that have the feature [−DELAYED_RELEASE], i.e. all stops. Indeed all languages in the PHOIBLE dataset have at least one stop.

```
(8) SELECT DISTINCT ?languages
    WHERE {?languages phoible:hasSegment ?segments .
    ?segments phoible:notHasFeature feature:DELAYED\_RELEASE }
```

Finally, the query in 9 checks the PHOIBLE data set for any languages that do not have a coronal phoneme.

```
(9) SELECT DISTINCT ?languages
    WHERE {?languages phoible:hasSegment ?segments .
    ?segments phoible:hasFeature feature:CORONAL }
```

The knowledge base, however, contains counter-evidence to the universal, found in the segment inventory of Northwest Mekeo [mek] (Jones, 1995, 1998), which has the consonants: / p, β, m, w, g, ŋ, j /. Northwest Mekeo's lack of coronals was reported in Blevins (2009), which was inspiration to compile a larger set of segment inventories than UPSID-451 and to develop a typological knowledge base using Linked Data; thus allowing users to query at the levels of segments and distinctive features to investigate questions regarding phonological typology and universals.

5 Conclusion

The knowledge base is a data-centric model. In comparison to individually devised relational databases, the knowledge base facilitates data sharing by publishing a self-describing data model according to explicitly encoded relationships found in the data. This graph data model is more dynamic and allows information to be added at any node. In my opinion, the graph data structure uses a technology that embraces principles towards a cyberinfrastructure approach (cf. Bender and Langendoen, 2010; Pericliev, 2010). The major benefit of using Linked Data is that the graph data structure is designed explicitly for data sharing. Because of global scope, the triple structure that makes up the graph allows for easy information integration. Two graphs from different sources that share a given URI can be merged without transforming the data. Another benefit, not explored in this paper, is the ability to mark-up RDF predicates with logical constructions using the Web Ontology Language (OWL, McGuinness and van Harmelen, 2004). For example, if a user does not consider length a phonemic property of segment inventories, then he or she can use the OWL property 'owl:sameAs' to state that the feature [+LONG] is equivalent to [−LONG]. With one simple statement the user can change the contents of the knowledge base, without changing the underlying data in the Linked Data graph.

Acknowledgements Thanks to Dan McCloy and Richard Wright for assisting me in mapping the SPA and UPSID segments into IPA and to Christopher Green for help with the collection and analysis of segment inventories from African languages. Thanks also to to three anonymous reviewers for feedback.

References

Beckett D (2004) RDF/XML Syntax Specification (Revised). Tech. rep., W3C, URL http://www.w3.org/TR/rdf-syntax-grammar/

Bender EM, Langendoen DT (2010) Computational Linguistics in Support of Linguistic Theory. Linguistic Issues in Language Technology (LiLT) 3(2):1–31

Berners-Lee T, Hendler J, Lassila O (2001) The Semantic Web. Scientific American URL http://www.sciam.com/article.cfm?articleID=00048144-10D2-1C70-84A9809EC588EF21

Bizer C, Cyganiak R, Heath T (2007) How to publish linked data on the web. http://www4.wiwiss.fu-berlin.de/bizer/pub/LinkedData Tutorial/

Blevins J (2009) Another Universal Bites the Dust: Northwest Mekeo Lacks Coronal Phonemes. Oceanic Linguistics 48(1):264–273

Cardoso J, Sheth AP (eds) (2006) Semantic Web Services, Processes and Applications, Springer

Chanard C (2006) Systèmes alphabétiques des langues africaines. http://sumale.vjf.cnrs.fr/phono/

Crothers JH, Lorentz JP, Sherman DA, Vihman MM (1979) Handbook of phonological data from a sample of the world's languages: A report of the stanford phonology archive

Hartell RL (ed) (1993) Alphabets des langues africaines. UNESCO and Société Internationale de Linguistique

Hayes B (2009) Introductory Phonology. Blackwell

Hyman LM (2008) Universals in phonology. The Linguistic Review 25:83–137

International Phonetic Association (2005) International Phonetic Alphabet. Tech. rep., International Phonetic Association, URL http://www.arts.gla.ac.uk/IPA/

Jones AA (1995) Mekeo. In: Tryon DT (ed) Comparative Austronesian Dictionary: An Introduction to Austronesian Studies, Part 1: Fascicle 2, Mouton de Gruyter

Jones AA (1998) Towards a Lexicogrammar of Mekeo (An Austronesian Language of Western Central Papua). Pacific Linguistics, Canberra

Lassila O, Swick RR (1999) Resource Description Framework (RDF): Model and syntax specification (recommendation). http://www.w3.org/TR/REC-rdf-syntax

Maddieson I (1984) Pattern of Sounds, Cambridge University Press, Cambridge, UK

Maddieson I, Precoda K (1990) Updating UPSID. In: UCLA Working Papers in Phonetics, vol 74, pp 104–111

McGuinness DL, van Harmelen F (2004) OWL Web Ontology Language Overview. URL http://www.w3.org/TR/owl-features/

Moran S (2012) Phonetics information base. PhD thesis, University of Washington

Pericliev V (2010) Machine-Aided Linguistic Discovery: An Introduction and Some Examples. London: Equinox

Prud'Hommeaux E, Seaborne A (2006) SPARQL query language for RDF. W3C working draft 4

The Unicode Consortium (2007) The Unicode Standard, Version 5.0.0, defined by: The Unicode Standard, Version 5.0. URL http://www.unicode.org/versions/Unicode5.0.0/

TYTO – A Collaborative Research Tool for Linked Linguistic Data

Andrea C. Schalley

Abstract In this paper, I introduce a computational tool, TYTO ("Typology Tool"), that utilises Semantic Web technologies in order to provide novel ways to process, integrate, and query cross-linguistic data. Its data store incorporates a set of ontologies (comprising linguistic examples, annotations, language background information, and metadata) backed by a logic reasoner software. This allows for highly targeted querying, and, with enough data on the relevant interest areas, TYTO can return answers to rather specific typological questions such as 'Which other languages in the North America, in addition to Yuchi, do encode senior kin and ingroup (such as belonging to the same ethnic group) in a suffixal case marking system?' TYTO's data store can be extended with additional ontologies and adapted to allow for project-specific analyses of linguistic data. It is further designed to facilitate collaboration and allow multi-user contributions, including automatic integration of data submitted at different stages by different contributors.

1 Introduction

What would, in an ideal world, an electronic resource supporting the division of typological theories in linguistics look like? The wish-list of typologists would presumably include the following:

1. *Cross-linguistic data*
   The resource should comprise data from all of the languages of the world, and this data should be comprehensive in order to allow for the cross-linguistic comparison of all form and meaning based differences found in the languages of the world. The data set should include both the raw data and its annotation or analysis by experts of the corresponding languages.

Andrea C. Schalley
Griffith University, Brisbane, Australia, e-mail: a.schalley@griffith.edu.au

C. Chiarcos et al. (eds.), *Linked Data in Linguistics*,
DOI 10.1007/978-3-642-28249-2_14, © Springer-Verlag Berlin Heidelberg 2012

2. *Grounding in actual linguistic examples*

It should provide access to actual linguistic examples for all analyses in order for theoretical claims to be verifiable. I.e. all data analyses contained in the resource should be documented and traceable back to example data, and via metadata information to the source of the example data.

3. *Data analysis*

Language data should be open for reanalysis, in order to correct and expand on previous analyses, and keep the overall system responsive to scientific progress. Such reanalyses should be tracked in the system, i.e. a history of sequential analyses of a data item should be accessible. Also, the data should be analysed very fine-grainedly, capturing information on all possible dimensions of typological variation.

4. *Querying and reporting*

The resource should be able to inform researchers in responding to highly targeted research questions such as 'Which other languages in the North America, in addition to Yuchi, do encode senior kin and in-group (such as belonging to the same ethnic or family group) in a suffixal case marking system?' Even more so, it should provide for extensive flexibility in accessing the data and their analyses. This in particular includes that a user-chosen number of variation dimensions should be dynamically combinable in user-defined queries. Such queries also determine in what way elements of the knowledge base should be presented, i.e. whether for instance just the number of languages that meet the query, a list of those languages, or a list of all examples for all of the languages should be reported. It should be self-explanatory to anyone how to formulate queries, and the user should be able to determine in which format the results of the query are presented.

5. *Scope*

In line with the aim of being able to compare linguistic form and meaning, the language data in the knowledge base should have been consistently analysed both with regard to their form and their meaning components. Going back to the above example, for a given data item the knowledge base should comprise whether this data item constitutes a suffix and/or a case marker, and whether it encodes the concepts KIN SENIOR and/or ETHNIC SAME. In other words, both a semasiological view and an onomasiological view on the language data should be facilitated by the knowledge base.

6. *Multi-user contributions*

The resource should allow multiple contributors and multiple 'consumers' (querying the knowledge base) to collaborate through it and use it at the same time. It should be able to handle overlapping contributions on the same and different areas of the knowledge base at the same and different times. Newly committed data should be automatically integrated into the knowledge base, and data should be made accessible for querying immediately after their submission.

7. *Fieldwork compatibility*

The resource should be deployable in the field, i.e. (i) it should be possible to store the whole resource locally on a computer and to run the system independently, without access to the Internet, (ii) researchers should be able to enter data collected in the field while in the field, and (iii) they should be able to query the system (including their own newly entered data) and obtain reports on the basis of the data while in the field, in order to inform their hypothesis forming and theory building processes. On their return, researchers should be able to easily and fast feed back their new data into the overall data store, where it should be integrated automatically.

8. *Data entry*

The entry of language data and their analyses should be organised as user-friendly as possible. It should meet the needs of typologists (e.g. in terms of automatically parsed interlinear glossing), be fast and efficient. The data entry should be flexible enough to provide interfaces for the submission of non-anticipated data be be entered.

9. *Expandability*

The knowledge base should be expandable. In particular, it should be possible to add new analytical concepts into a structured interconnected 'vocabulary' of descriptive and semantic concepts. Terminological controversies should be catered for, by either resolving them successfully or by integrating all options into the 'vocabulary'. Also, the knowledge base should ideally be able to hold information on both what is possible and what is not possible in a particular language or in the languages across the world more generally (i.e. it should contain positive and negative evidence).

Of course, we do not live in an ideal world, and there are obvious roadblocks to implementing the wish-list fully, and the most obvious ones amongst them are: (i) We do not have enough information gathered about all languages in the world, and hence comprehensiveness cannot be achieved. (ii) Related to this, for an analysis of each language with regard to all typological variation dimensions, we would have to have a complete list of those variation dimensions. This is not the case to date, and much debate is ensuing about those dimensions and the question of language universals more generally (cf., e.g. the recent discussion about the 'myth of language universals', Evans and Levinson, 2009; Rooryck et al., 2010). (iii) Even if (i) and (ii) were solvable, the amount of expert analysis that would have to go into the knowledge base would be immeasurable, and a never-ending task if one takes language change into account. (iv) Terminological controversies are unlikely to be successfully resolved, and it has to be acknowledged that there is a terminological tension between the description of a single language and the comparison of several or all languages (cf., e.g., the discussion in Corbett, 2007; Haspelmath, 2010).

Despite thus entertaining an unrealistic vision, we can try to come as closely as possible to this ideal – with the current computational resources we have available. In this paper, I introduce a typology tool, TYTO, that has been designed to implement as many of these desiderata as possible. While not fully completed to date, I will give an overview of what the tool can and cannot do as well as outline some of the design decisions and technologies it is based upon in Sect. 3. Before going into detail about TYTO, I will very briefly look at predecessing and related current systems in Sect. 2.[1] I will conclude with an outlook on the further development of TYTO in Sect. 4.

2 Related Projects

One of the first projects that aimed at creating an online resource for typological information was the *Cross-linguistic Reference Grammar (CRG)*, which was initiated jointly by Bernard Comrie, William Croft, Christian Lehmann, and Dietmar Zaefferer about two decades ago (Comrie et al., 1993; Zaefferer, 2006). It aimed "to create some kind of revised electronic version of the famous Lingua descriptive studies [LDS, AS] questionnaire" (Zaefferer, 2006, p. 113), as published in Comrie and Smith (1977), and hence at integrated comprehensive grammatical descriptions of languages. CRG's knowledge base consists of a predefined AND-OR-tree.

In terms of current systems, one of the prominent ones surely is *The World Atlas of Language Structures (WALS)*, which "is a large database of structural (phonological, grammatical, lexical) properties of languages gathered from descriptive materials (such as reference grammars) by a team of 55 authors (many of them the leading authorities on the subject)" (Dryer and Haspelmath, 2011). WALS, in its current second version, holds an impressive amount of cross-linguistic information. It comprises 192 features such as 'Voicing in Plosives and Fricatives', 'Sex-based and Non-sex-based Gender Systems', 'Position of Case Affixes', 'SOVNeg Order', or 'Tea', and hence contains both semasiologically-oriented as well as onomasiologically-oriented features. Contributions are submitted by experts on the features (and not by experts on the languages). It allows to cross-search for two properties from its public interface.

The *Database of Syntactic Structures of the World's Languages (SSWL)* also allows a cross-search for two properties from its public interface. However, in contrast to WALS, contributors are not experts on the properties but experts but on the respective language on which they contribute. "SSWL is a searchable database that allows users to discover which properties (morphological, syntactic, and semantic) characterize a language, as well as how these properties relate across languages. This system is designed to be free to the public and open-ended. Anyone can use the database to perform queries." (SSWL, n.d.) It encompasses data on 130 languages and 54 binary properties, and has received language analyses from 202 contributors to date (SSWL, n.d., as of 27 Nov 2011).

While WALS and SSWL are typological databases in themselves, the from a technical viewpoint closest system to TYTO is the *Typological Database System (TDS)*, which "is a web-based service that provides integrated access to a collection

---

[1] For a still brief, but more explicit comparison of TYTO with some of the approaches listed in Sect. 2 (also regarding some technical features), cf. Borkowski and Schalley (2011).

of independently created typological databases." (Dimitriadis et al., 2009, p. 155) It thus provides an interface to the data contained in other typological databases. It supports unified querying across these typological resources with the help of an integrated ontology (network of cross-connected concepts relevant for the domain), and uses a bottom-up approach to the development of this ontology.

The last related resource to be listed here is the grammar-authoring tool *Galoes* (Galoes, n.d.; Nordhoff, 2008). Currently, there is one full grammatical description available, and about another six partial descriptions exist (although some do not allow read access yet). "GALOES supports rendering of linguistic examples, embedded audio files, cross-references, collaboration tools and more. There is an online version on www.galoes.org/grammars and an offline client to be used on your laptop." (Galoes, n.d.) The database can be searched, either by choosing the free search function or pre-defined question from, e.g., the LDS questionnaire or WALS.

3 TYTO

Yet, typologists will continue to encounter specific questions for which answers or at least leads to answers cannot be readily obtained from the resources outlined above – the array of conceivable questions seems limitless. The resource introduced in the following, TYTO,[2] is intended to provide more flexibility in the kind of questions that can be posed to the system, and this of course has implications on how the underlying knowledge base is set up, and what technologies are used to achieve this responsiveness to the needs of users.

In the following, I will take up the wish-list from Sect. 1, addressing in each case how TYTO fares with regard to these ideals. Given the space limitations in this chapter, I will neither be able to provide very detailed accounts of the plethora of topics touched upon in the wish-list nor explain TYTO's technical solutions in detail. This chapter is merely intended as an introduction to the TYTO tool and its underlying ideas. For additional information on some of the approaches taken and design decisions made, cf. Borkowski and Schalley (2011) and Schalley (in press).

3.1 Cross-Linguistic Data

As indicated above, this ideal will remain wishful thinking, given the scattered knowledge we have about the languages of the world, the incomplete list of vari-

---

[2] TYTO is being developed as part of the Australian Research Council Discovery project DP0878126 ("Social cognition and language"). The aims of this project include the building of a detailed and cross-linguistically valid model of how social cognition is grammaticalized across the world's languages. This is approached through the systematization and synthesis of already-recorded material ('library sample') and newly-gathered data for a small number of languages ('fieldwork data'). TYTO is the tool that will eventually provide the infrastructure to move this model into an electronic, dynamically queryable, and extensible format.

ation dimensions across languages, and our limited resources to describe more languages – or already described languages in more detail (i.e. gather more language data, and analyze it fine-grainedly). Nonetheless, whatever is collected can in principle be integrated into TYTO. However, to attain the aim of a high comparability of cross-linguistic data and hence flexible searchability across the underlying knowledge base, the information has to be semantically tagged in a consistent way. This is achieved using Semantic Web concepts. An underlying ontology forms the backbone of TYTO's knowledge base; it consists of a set of interrelated sub-ontologies, including: (a) the primary semantic domain (in our project: social cognition); (b) linguistic examples (both in their original form and potentially revisions of this original); (c) linguistic annotations (cross-linguistic description with respect to both form and function); (d) language background information (family, size, vitality, geographic region [linguistic and political], society [economy, religion, tradition, etc.]); and (e) metadata (example source information [fieldwork, literature]; general metadata [contributor information etc.]).

The ontology is being developed using the Web Ontology Language (OWL, McGuinness and van Harmelen, 2004) and the ontology editor Protégé (BMIR, 2011). Due to the modular organization of the overall ontology into these sub-ontologies, reusability of each of the overall ontology's parts is ensured and encouraged. It is envisaged that some of the sub-ontologies will in the future be linked to corresponding resources such as Ethnologue (Lewis, 2009).

3.2 Grounding in Actual Linguistic Examples

TYTO's development is data-driven. The ontology and hence the knowledge base will be incrementally built up, through the analyses that are entered for linguistic examples. The linguistic examples themselves will also be stored in the knowledge base, "in a self-contained XML data fragment whose structure is based on the general model for interlinear text proposed by Bow et al. (2003) and specified by a separate XML schema." (Borkowski and Schalley, 2011) All data analyses are thus verifiable and can be traced back to specific linguistic examples. We believe that only a data-driven approach will generate the fine-grainedness of ontological concepts, relations, and constraints needed for typological work. This approach stands in sharp contrast to the approach that has been adopted for the development of GOLD, the *Generalized Ontology for Linguistic Description* (GOLD, 2010).[3]

3.3 Data Analysis

TYTO offers the option to revise example analyses and keeps a history of these revisions in the running system, i.e. this history can be directly accessed through

---

[3] For critical comments on GOLD, cf. Cysouw et al. (2005) and Munro and Nathan (2005).

queries. In principle, it is possible to analyse each linguistic example with regard to all possible variation dimensions (i.e. those that are relevant to it). This is unlikely to happen in any one analysis step, so allowing for revisions is a way of also allowing the addition of analyses with regard to additional variation dimensions (the same linguistic example can hence function as the evidence for several or all variation dimensions).

3.4 Querying and Reporting

Users are able to pose targeted questions as required by their research foci and needs, via tailored queries. The system is based on a logic reasoner software and Semantic Web technology (SPARQL, Prud'hommeaux and Seaborne, 2008) and hence can be queried in a flexible way that allows for combining a chosen number of variation dimensions comprised in the ontology within a query. Provided with the right query, TYTO can answer highly targeted questions such as our example from Sect. 1. What and how information is presented is specified through the reporting engine (Jasper-Reports, JasperForge, 2000-2010), which also allows for a wide range of output formats (including, e.g., PDF). The only current drawback is that the user needs to somewhat familiarise themselves with the structure of the ontology, the query language, and the report designer. However, designed reports can be shared and reused easily in the user community.

3.5 Scope

As indicated previously, an analysis both with regard to form and meaning components is possible using TYTO, i.e. both a semasiological view and an onomasiological view on the language is facilitated (and made explicit in the corresponding substructuring of the ontology). It is extremely unlikely that analysis comprehensiveness will be achievable (given knowledge and resource limitations), but this is catered for by the open-world assumption underlying TYTO's ontological approach: if the reasoner cannot prove a statement to be true, such a statement being considered as unknown, and the system can deal with this lack of knowledge.[4] The open-world assumption thus appropriately reflects the fact that neither single contributors or users of the knowledge base nor the overall knowledge base will ever have complete knowledge.

---

[4] Even though this might sound very natural, it is in contrast to most computational approaches that use the closed-world assumption and hence assume that their knowledge about the application domain is complete.

3.6 Multi-User Contributions

The tool is specifically designed as to support collaborative effort in linguistic typology, offering the potential of flexible contributions spanning across different times, locations, and groups of contributors and 'consumers'. This is achieved through an elaborate version control system, which will run automatic consistency checks on the ontology using the reasoner, for automatic integration of data into the knowledge base (or human postprocessing in the case of conflicts). For a more detailed description of this process, including how TYTO deals with conflicting analyses by contributors and how it is envisaged to persuade a significant number of potential contributors to participate (via creating a critical mass of data at the outset), cf. Borkowski and Schalley (2011) and Schalley (in press).

3.7 Fieldwork Compatibility

The software is designed to be installed locally on users' computers. "This permits use while on fieldwork – disconnected from the Internet – [...], so interested parties can use the complete TYTO system independently" (Borkowski and Schalley, 2011). Fieldwork results can then be easily fed back into the central TYTO knowledge base; this will be handled by the version control system, as this is a special case of employing the multi-contributor capabilities.

3.8 Data Entry

An initial input system has been developed which takes interlinear glossed data, parses it on the basis of the Leipzig Glossing Rules (Bickel et al., 2008; with some additions as necessary for computational processing). Each component is then available for selection and can be linked to categories in the ontology, e.g. a suffix indicating that the denoted entity is a member of the ethnic in-group can be linked to the ontology category ETHNIC SAME. In addition, form fields are available, for entering metadata and other relevant information, such as the language (e.g. Yuchi), for which the data is an example. These form fields also provide an option for entering non-anticipated data. For a more detailed description of the workflow, cf. Borkowski and Schalley (2011) and Schalley (in press).

3.9 Expandability

TYTO's ontological approach, with its open-world assumption, comes with the advantages of ontologies: ease of knowledge base extension and maintainability. It is possible to integrate new concepts into the ontology (the knowledge base's structured interconnected 'vocabulary' of descriptive and semantic concepts), change the ontological class hierarchy (including multiple inheritance), add additional linguistic examples, and provide further analyses to already existing linguistic examples. TYTO, given its data-driven nature, can, however, only provide positive evidence at this stage in its development, as an analysis of linguistic examples that are well-formed can only provide information on what is possible.

4 Conclusion

This chapter could only provide a rather cursory introduction to the collaborative typology tool TYTO. TYTO will remain a major enterprise requiring further concerted development to bring all its features to full fruition. The design of the tool is very modular, it consists of a number of separable components that other projects can select and adjust to their needs by mixing and matching as required. The incremental development of the ontology will continue for years to come, and roadblocks are expected in particular for the ontology development due to terminological controversies and some inadequate descriptive devices for a number of linguistic phenomena. Nonetheless, it is worthwhile pursuing this path, as it will bring issues out into the open as well as collate information. We are planning on linking this information to resources holding related information about, e.g., language, culture, and geography, as well as making our own data available in linkable form.

Acknowledgements I gratefully acknowledge the support I received from the Australian Research Council (Grant *Social Cognition and Language*, DP0878126). In addition, I would in particular like to thank Alexander Borkowski and Nicholas Evans for their collaboration on this project. Without our stimulating discussions, their insight, and their invaluable contributions on the conceptualization, design, and implementation of TYTO, the tool would not be where it is now.

References

Bickel B, Comrie B, Haspelmath M (2008) Leipzig glossing rules: Conventions for interlinear morpheme-by-morpheme glosses. Available online at http://www.eva.mpg.de/lingua/resources/glossing-rules.php. Accessed on 2011-11-27

BMIR (2011) The Protégé Ontology Editor and Knowledge Acquisition System. Stanford Center for Biomedical Informatics Research. Available online at http://protege.stanford.edu/. Accessed on 2011-11-27

Borkowski A, Schalley A (2011) Going beyond archiving - a collaborative tool for typological research. In: Thieberger N, Barwick L, Billington R, Vaughan J (eds) Sustainable data from digital research: Humanities perspectives on digital scholarship, Custom Book Centre, University of Melbourne, Melbourne

Bow C, Huges B, Bird S (2003) Towards a general model of interlinear text. In: Proceedings of EMELD 2003, available online at http://emeld.org/workshop/2003/bowbadenbird-paper.pdf. Accessed on 2011-11-27

Comrie B, Smith N (1977) Lingua descriptive studies: questionnaire. Lingua 42:1–72

Comrie B, Croft W, Lehmann C, Zaefferer D (1993) A framework for descriptive grammars. In: Crochetière A, Boulanger JC, Ouellon C (eds) Actes du XVe Congrès International des Linguistes/Proceedings of the XVth International Congress of Linguists, Les Presses de l'Université Laval, Sainte-Foy, pp 159–170

Corbett G (2007) Canonical typology, suppletion, and possible words. Language 83(1):8–42

Cysouw M, Good J, Albu M, Bibiko HJ (2005) Can gold 'cope' with wals? retrofitting an ontology onto the world atlas of languages structures. In: Proceedings of EMELD 2005. Available online at http://emeld.org/workshop/2005/proceeding.html. Accessed on 2011-11-27

Dimitriadis A, Windhouwer M, Saulwick A, Goedemans R, Bíró T (2009) How to integrate databases without starting a typology war: the typological database system. In: Everaert M, Musgrave S, Dimitriadis A (eds) The Use of Databases in Cross-Linguistic Studies, Mouton de Gruyter, Berlin, pp 155–207

Dryer M, Haspelmath M (eds) (2011) The World Atlas of Language Structures Online. Max Planck Digital Library, Munich, available online at http://wals.info/. Accessed on 2011-11-27.

Evans N, Levinson S (2009) The myth of language universals: Language diversity and its importance for cognitive science. Behavioral and Brain Sciences 32:429–492

Galoes (n.d.) Available online at http://www.galoes.org/. Accessed on 2011-11-27.

GOLD (2010) Generalised Ontology for Linguistic Description. Available online at http://www.linguistics-ontology.org/gold.html. Accessed on 2011-11-27.

Haspelmath M (2010) Comparative concepts and descriptive categories in crosslinguistic studies. Language 86(3):663–687

JasperForge (2000-2010) Jasperreports: Open Source Java Reporting Library. Available online at http://jasperforge.org/projects/jasperreports. Accessed on 2011-11-27.

Lewis M (ed) (2009) Ethnologue: Languages of the World, Sixteenth edition. SIL International, Dallas, online version available at http://www.ethnologue.com/. Accessed on 2011-11-27.

McGuinness D, van Harmelen F (2004) Owl web ontology language. overview. W3C Recommendation 10 February, available online at http://www.w3.org/TR/owl-features/. Accessed on 2011-11-27

Munro R, Nathan D (2005) Towards portability and interoperability for linguistic annotation and language-specific ontologies. In: Proceedings of EMELD 2005, available online at http://emeld.org/workshop/2005/proceeding.html. Accessed on 2011-11-27

Nordhoff S (2008) Electronic reference grammars for typology: challenges & solutions. Language Documentation and Conservation 2(2):296–324

Prud'hommeaux E, Seaborne A (2008) SPARQL Query Language for RDF. W3C Recommendation 15 January, available online at http://www.w3.org/TR/rdf-sparql-query/. Accessed on 2011-11-27.

Rooryck J, Smith N, Liptak A, Blakemore editors D (2010) Special issue on Evans & Levinson's "The myth of language universals". Lingua 120(12):2651–2758

Schalley A (in press) Many languages, one knowledge base: Introducing a collaborative ontolinguistic research tool. In: Schalley A (ed) Practical Theories and Empirical Practice, John Benjamins, Amsterdam/Philadelphia

SSWL (n.d.) Database of Syntactic Structures of the World's Languages. Available online at http://sswl.railsplayground.net/. Accessed on 2011-11-27.

Zaefferer D (2006) Realizing Humboldt's dream: Cross-linguistic grammatography as data-base creation. In: Ameka F, Dench A, Evans N (eds) Catching Language: The Standing Challenge of Grammar-Writing, Mouton de Gruyter, Berlin, pp 113–136

The Open Linguistics Working Group of the Open Knowledge Foundation

Christian Chiarcos, Sebastian Hellmann, and Sebastian Nordhoff

Abstract The Open Linguistics Working Group (OWLG) is an initiative of experts from different fields concerned with linguistic data, including academic linguistics (e.g. typology, corpus linguistics), applied linguistics (e.g. computational linguistics, lexicography and language documentation) and NLP (e.g. from the Semantic Web community). The primary goals of the working group are 1) the promotion of the idea of open linguistic resources 2) the development of means for their representation, and 3) encouraging the exchange of ideas across different disciplines.

To a certain extent, the activities of the Open Linguistics Working Group converge towards the creation of a Linguistic Linked Open Data cloud, which is a topic addressed from different angles by several members. Some of these activities are described further in the other contributions of this part.

1 The Open Knowledge Foundation

The Open Knowledge Foundation (OKFN) is a not-for-profit organization founded in 2004 for the promotion of open knowledge, i.e. any kind of data and content that can be freely used, reused, and redistributed. Activities of the OKFN include the

Christian Chiarcos
Information Sciences Institute, University of Southern California, 4676 Admiralty Way # 1001, Marina del Rey, CA 90292 e-mail: chiarcos@daad-alumni.de

Sebastian Hellmann
Universität Leipzig, Fakultät für Mathematik und Informatik, Abt. Betriebliche Informationssysteme, Johannisgasse 26, 04103 Leipzig, Germany e-mail: hellmann@informatik.uni-leipzig.de

Sebastian Nordhoff
Department of Linguistics, Max Planck Institute for Evolutionary Anthropology, Deutscher Platz 6, 04103 Leipzig, Germany e-mail: sebastian_nordhoff@eva.mpg.de

C. Chiarcos et al. (eds.), *Linked Data in Linguistics*,
DOI 10.1007/978-3-642-28249-2_15, © Springer-Verlag Berlin Heidelberg 2012

development of standards (Open Definition), tools (CKAN) and support for working groups and events:

The **Open Definition** sets out principles to define 'openness' in relation to content and data.[1] This definition can be summed up in the statement that "A piece of content or data is open if anyone is free to use, reuse, and redistribute it – subject only, at most, to the requirement to attribute and share-alike."

The OKFN provides a catalog system for open datasets, **CKAN**.[2] CKAN is an open-source data portal software developed to publish, to find, and to reuse open content and data easily, especially in ways that are machine automatable.

The OKFN also serves as host for various working groups addressing problems of open data in different domains. At the time of writing, there are 18 OKFN **working groups** covering fields ranging from government data and economics over archeology or open text books to cultural heritage.[3] The OKFN organizes various events such as the Open Knowledge Conference (OKCon) and facilitates the communication between different working groups.

In 2010, the **OKFN Working Group on Open Linguistic Data** (OWLG) was founded. Since its formation, the Open Linguistics Working Group has been steadily growing. We have identified goals and problems that are to be addressed, and directions that are to be pursued in the future. Preliminary results of this on-going discussion process are summarized in this contribution: Section 2 specifies the goals of the working group by identifying seven points that emerged from our discussions; Sect. 3 identifies four major problems and challenges of the work with linguistic data; and, finally, Sect. 4 gives an overview of recent activities and the current status of the group.

2 Goals of the Open Linguistics Working Group

As a result of numerous discussions with interested linguists, NLP engineers and information technology experts, we identified seven open problems for our respective communities and their ways to use, to access and to share linguistic data. These represent the challenges to be addresses by the working group, and the role that it is going to fulfill:

1. Promote the idea of open data in linguistics and in relation to language data.
2. Act as a central point of reference and support for people interested in open linguistic data.
3. Provide guidance on legal issues surrounding linguistic data to the community.
4. Build an index of indexes of open linguistic data sources and tools and link existing resources.
5. Facilitate communication between existing groups.

---

[1] http://www.opendefinition.org

[2] http://ckan.org/

[3] For a complete overview see http://okfn.org/wg.

6. Serve as a mediator between providers and users of technical infrastructure.
7. Assemble best-practice guidelines and use cases to create, use and distribute data.

In many aspects, the OWLG is not unique with respect to these goals. Indeed, there are numerous initiatives with similar motivation and overlapping goals, e.g. the Cyberling blog,[4] the ACL Special Interest Group for Annotation (SIGANN),[5] and large multi-national initiatives such as the ISO initiative on Language Resources Management (ISO TC37/SC4),[6] the American initiative on Sustainable Interoperability of Language Technology (SILT),[7] or European projects such as the initiative on Common Language Resources and Technology Infrastructure (CLARIN),[8] the Fostering Language Resources Network (FLaReNet),[9] and the Multilingual Europe Technology Alliance (META).[10]

The key difference between these and the OWLG is that we are not grounded within a *single* community, or even restricted to a hand-picked set of collaborating partners, but that our members represent the whole band-width from academic linguistics (with its various subfields, e.g. typology and corpus linguistics) over applied linguistics (e.g. language documentation, computational linguistics, computational lexicography) and computational philology to Natural Language Processing and information technology. We do not consider ourselves to be in competition with any existing organization or initiative, but we hope to establish new links and further synergies between these. Section 3 summarizes typical and concrete scenarios where such an interdisciplinary community may help to resolve problems observed (or, sometimes, overlooked) in the daily practice of working with linguistic resources.

3 Open Linguistics Resources, Problems and Challenges

Among the broad range of problems associated with linguistic resources, we identified four major classes of problems and challenges that may be addressed by the OWLG:

legal questions   There is a great uncertainty with respect to legal questions of the creation and distribution of linguistic data.
technical problems   Often, researchers come up with questions regarding the choice of tools, representation formats and metadata standards for different types of linguistic annotation.

---

[4] http://cyberling.org/
[5] http://www.cs.vassar.edu/sigann/
[6] http://www.tc37sc4.org
[7] http://www.anc.org/SILT
[8] http://www.clarin.eu
[9] http://www.flarenet.eu
[10] http://www.meta-net.eu

repository of open linguistic resources   So far, the communities involved have not yet established a common point of reference for existing open linguistic resources; at the moment there are multiple metadata collections.

spread the word   Finally, there is an agitation challenge for open data in linguistics, i.e. how (and whether) we should convince our collaborators to release their data under open licenses.

3.1 Legal Questions

The linguistic community is increasingly becoming aware of the potentially difficult legal status of different types of linguistic resources. Typical questions include: *How to find a suitable license for my corpus?*, *Whose copyright do I have to respect?* (for example, corpora may have complex copyright situations where the original authors own the primary data, and thus may have *partial* copyright on the entire collection), *Are there exceptions (e.g. for academic research) to the copyright that may allow me to work with my corpus anyway?*, *How to circumvent (or solve) copyright issues?*, *What legal restrictions apply to a particular resource (e.g. web corpora, newspaper corpora, digitizations of printed editions, audio and video files) ?*, or *How to create multi-media (audio, video) data collections in a way that allows us to use (and hopefully, distribute) them for research?*

The situation is even more complex because the legal situation may change over time, and this complexity multiplies on an international scale. The OWLG can provide a platform to discuss such problems, to collect recommendations and document use cases as found in publications and technical reports, and discussed on conferences and mailing lists.

3.2 Technical Problems

When creating a new corpus in a novel domain, researchers are confronted with the question which tool to choose for which type of annotation. The OWLG can collect case studies and best practice recommendations with respect to this, it will encourage the documentation of use cases, collect links to documented case studies and best practice recommendations (e.g. by the American project on Electronic Metadata for Endangered Languages Data (EMELD, 2002-2007),[11] or FLaReNet),[12] and participate in the maintenance of existing sites that provide an overview of annota-

---

[11] http://emeld.org/school

[12] http://www.flarenet.eu/?q=Standards_and_Best_Practices

tion tools and their domains of application (e.g. the Linguistic Annotation Wiki,[13] or corresponding parts of the Wiki of the Association of Computational Linguistics).[14]

A question related to the choice of tools is the choice of representation formalisms. Interoperability between different representation formalisms used in NLP pipelines, linguistic corpora and other types of linguistic resources has become a major field of research in the last years. We intend to provide basic information about proposed standard formats – e.g. the ISO TC37/SC4 proposal LAF/GrAF (Ide and Suderman, 2007), the specifications of the Text Encoding Initiative (Ide and Veronis, 1995, TEI) – and applicable formalisms (e.g. XML or RDF). In this volume, several publications addressed interoperability issues and possible solutions based on RDF and related formalisms.

These formats, again, are closely related to the question which corpus infrastructure (data base, search interface) may be suitable to store, query and visualize what kind of linguistic annotations, e.g. domain- and community-specific tools like Toolbox (Busemann and Busemann, 2008) and ELAN (Hellwig et al., 2008), or general-purpose corpus query tools like ANNIS (Zeldes et al., 2009). A third problem is the question of documentation requirements for different types of resources, the use of metadata standards (e.g. Dublin Core, Weibel et al., 1998, or the TEI header, Giordano, 1995), and how annotation documentation and interoperability can be improved by linking linguistic resources with terminology repositories (e.g. the General Ontology of Linguistic Description GOLD, Farrar and Langendoen, 2003, or the ISO TC37/SC4 Data Category Registry, Kemps-Snijders et al., 2009). The OWLG aims to collect such questions and (partial) answers to these. We will contribute to existing metadata repositories and cooperate with other initiatives that pursue similar goals, e.g. the ACL Special Interest Group in Linguistic Annotation (SIGANN).[15]

Within the working group, we encourage (but do not require) the conversion of linguistic resources to Linked Data,[16] and selected activities in this direction currently conducted by different OWLG members are described here by Chiarcos (this vol.), Hellmann et al. (this vol.), Nordhoff (this vol.) and Chiarcos et al. (this vol.) who discuss formalisms and technologies for the modeling and the interlinking of representative types of linguistic data in a Linguistic Linked Open Data cloud.

---

[13] http://annotation.exmaralda.org/index.php/Linguistic_Annotation
[14] http://aclweb.org/aclwiki/index.php?title=Tools_and_Software_for_English
[15] http://www.cs.vassar.edu/sigann
[16] http://linkeddata.org

3.3 Overview over Existing Resources

If a new research question is to be addressed, the question arises which resources
may already be available and whether these may be accessible. Often, this problem
is still solved by asking experts on mailing lists.[17]

In order to establish a more structured way of collecting information, the OWLG
has begun to collect metadata about open linguistic resources within the CKAN
repository.[18] CKAN is qualitatively different from earlier metadata repositories[19]
in two respects:

- CKAN focuses on the license status of the resources and it encourages the use of
  open licenses.
- CKAN is **not restricted to linguistic resources**, but rather, it is used by all working groups, as well as interested individuals outside these working groups. Examples for CKAN resources created outside the linguistic/NLP community that
  are nevertheless of relevance to linguists include collections of open textbooks,[20]
  the complete works of Shakespeare,[21] or the Open Richly Annotated Cuneiform
  Corpus (ORACC).[22]

3.4 Agitation

One of the goals of the OWLG is the promotion of open licenses for linguistic data
collections. As we know from practical experience, researchers sometimes hesitate
to provide their data under an open license. This is partially due to the uncertainty
with respect to the legal situation, but there are also certain **sociological** factors, e.g.
the (understandable) fear that people exploit the resources before the original author
had the chance to do so.

We hope to encourage the discussion of legal issues and to provide case studies
that may help clarify these problems. For example, a solution for the sociological
aspect mentioned above may be that data collections are designed as open linguistic
resources from the beginning, but that their publication is delayed for several years,
so that the creators can make private use of their data long enough before any con-
current may get hands on it. One important argument that favors the use of open
resources in academia is that only resources that are available to other researchers

---

[17] For example, the CORPORA mailing list, http://listserv.linguistlist.org/
archives/corpora.html

[18] http://ckan.net, resources tagged by linguistics.

[19] For example, those maintained by META-NET (http://www.meta-net.eu)
FLaReNet (http://www.flarenet.eu/?q=Documentation_about_Individual_
Resources) or CLARIN (http://catalog.clarin.eu/ds/vlo.)

[20] http://wiki.okfn.org/Wg/opentextbooks

[21] http://openshakespeare.org

[22] http://oracc.museum.upenn.edu

make it possible that linguists working empirically meet elementary scientific standards such as verifiability.

4 Recent Activities and Ongoing Developments

In the first year of its existence, the OWLG focused on the questions to address, the formulation of general goals, and the identification of potentially fruitful application scenarios. At the time of writing, we have reached a critical step in the formation process of the working group: Having defined a (preliminary) set of goals and principles, we can now concentrate on the tasks at hand, e.g. to collect resources and to attract interested people in order to address the challenges identified above.

As of October, 12th, 2011, the Working Group assembles 67 people from 29 different organizations and 10 countries.[23] Our group is relatively small, but continuously growing and sufficiently heterogeneous. It includes people from library science, typology, historical linguistics, cognitive science, computational linguistics, and information technology, just to name a few, so, the ground for fruitful interdisciplinary discussions has been laid out.

The Working Group maintains a home page,[24] a mailing list,[25] a wiki,[26] and a (guest) blog,[27] currently featuring contributions by Nancy Ide (Text Encoding Initiative, American National Corpus, Vassar College) and Christiane Fellbaum (Word-Net, University of Princeton). We conduct regular meetings and have established a series of workshops (Workshop on Open Data in Linguistics at the 6th Open Knowledge Conference (OKCon 2011, Berlin, Germany, June 30th, 2011); Workshop on Linked Data in Linguistics (LDL 2012, Frankfurt/M., Germany, March 7th-9th)).

Recent community activities include the compilation of a list of resources that represent interesting candidates for the Linguistic Linked Open Data cloud. Most of these resources are free, others are partially free (i.e. annotations free, but text under copyright), and a few have been included that are not free, but very representative for a particular type of resource (e.g. corpora derived from the Penn Treebank (Marcus et al., 1994) as prototypical examples for multi-layer corpora). As of Dec, 22th, 2011, this list comprises 103 resources, including lexicons, word lists, corpora, and collections of linguistic metadata. Subsequently, these resources are registered at the CKAN metadata repository and a few will be selected for deeper investigation. The development of a collection of open, freely accessible linguistic resources that are represented in interoperable standards represents a concrete goal for several members of the working group and may be seen as a long-term vision of the OWLG.

---

[23] Germany, US, UK, France, Canada, Australia, the Netherlands, Greece, Hungary, Slovenia

[24] http://linguistics.okfn.org

[25] http://lists.okfn.org/mailman/listinfo/open-linguistics

[26] http://wiki.okfn.org/Wg/linguistics

[27] http://blog.okfn.org/category/working-groups/wg-linguistics

The identification of these resources will be followed by the analysis of possible links between them, and the generation of a Linguistic Linked Data Cloud. A summary of first steps into this direction are described by Chiarcos et al. (this vol.).

References

Busemann A, Busemann K (2008) Toolbox self-training. Tech. rep., http://www.sil.org, version 1.5.4, Oct 2008

Chiarcos C (this vol.) Interoperability of corpora and annotations. pp 161–179

Chiarcos C, Hellmann S, Nordhoff S (this vol.) Linking linguistic resources: Examples from the Open Linguistics Working Group. pp 201–216

Farrar S, Langendoen DT (2003) A Linguistic Ontology for the Semantic Web. GLOT International 7:97–100

Giordano R (1995) The TEI header and the documentation of electronic texts. Computers and the Humanities 29(1):75–84

Hellmann S, Stadler C, Lehmann J (this vol.) The German DBpedia: A sense repository for linking entities. pp 181–189

Hellwig B, Uytvanck DV, Hulsbosch M (2008) ELAN - Linguistic Annotator. Tech. rep., http://www.lat-mpi.eu/tools/elan, version of 2008-07-31

Ide N, Suderman K (2007) GrAF: A graph-based format for linguistic annotations. In: Proc. Linguistic Annotation Workshop (LAW 2007), Prague, Czech Republic, pp 1–8

Ide N, Veronis J (1995) Text encoding initiative: Background and contexts. Kluwer Academic Pub

Kemps-Snijders M, Windhouwer M, Wittenburg P, Wright S (2009) ISOcat: Remodelling metadata for language resources. International Journal of Metadata, Semantics and Ontologies 4(4):261–276

Marcus M, Santorini B, Marcinkiewicz M (1994) Building a large annotated corpus of English: The Penn Treebank. Computational Linguistics 19(2):313–330

Nordhoff S (this vol.) Linked Data for linguistic diversity research: Glottolog/Langdoc and ASJP. pp 191–200

Weibel S, Kunze J, Lagoze C, Wolf M (1998) RFC 2413 - Dublin Core metadata for resource discovery. http://www.isi.edu/in-notes/rfc2413.txt

Zeldes A, Ritz J, Lüdeling A, Chiarcos C (2009) ANNIS: A search tool for multi-layer annotated corpora. In: Proc. Corpus Linguistics, Liverpool, UK, pp 20–23

Interoperability of Corpora and Annotations

Christian Chiarcos

Abstract This paper describes the application of OWL and RDF to address the interoperability of linguistic corpora and linguistic annotations within such corpora. Interoperability of linguistic corpora involves two aspects: Structural interoperability (annotations of different origin are represented using the same formalism) and conceptual interoperability (annotations of different origin are linked to a common vocabulary).

Building on an infrastructure developed to represent, to store, to query and to visualize multi-layer corpora with any kind of text-oriented annotation (Chiarcos et al., 2008), this paper proposes to address both aspects by means of OWL/RDF-based formalisms. Key advantages of this approach include the existence of a rich technological ecosystem developed around RDF and OWL, the conceptual similarity of generic data models for linguistic annotations and RDF (both based on labeled directed graphs), and the application of OWL/DL reasoners that can be applied to validate the consistency of linguistic corpora and their annotations and to infer additional information that is relevant, for example, for their appropriate visualization.

Additionally, representing corpora in OWL and RDF allows to interlink resources freely, e.g., different annotation layers of a multi-layer corpus, translated texts in parallel corpora, or linguistic corpora and lexical-semantic resources. Modeled in this way, corpora can be fully integrated in a Linked Open Data (sub-)cloud of linguistic resources, along with lexical-semantic resources and knowledge bases of information about languages and linguistic terminology.

Christian Chiarcos
Information Sciences Institute, University of Southern California, 4676 Admiralty Way # 1001, Marina del Rey, CA 90292 e-mail: chiarcos@daad-alumni.de

C. Chiarcos et al. (eds.), *Linked Data in Linguistics*,
DOI 10.1007/978-3-642-28249-2_16, © Springer-Verlag Berlin Heidelberg 2012

1 Motivation and Background

In recent years, the interoperability of linguistic resources has become a major topic in the fields of computational linguistics and Natural Language Processing (Ide and Pustejovsky, 2010): Within the last 30 years, the maturation of language technology and the increasing importance of corpora in linguistic research produced a growing number of linguistic corpora with increasingly diverse annotations. While the earliest annotations focused mostly on part-of-speech and syntax annotation, later NLP research included also on semantic, anaphoric and discourse annotations, and with the rise of statistic MT, a large number of parallel corpora became available. In parallel, specialized technologies were developed to represent these annotations, to perform the annotation task, to query and to visualize them. Yet, the tools and representation formalisms applied were often specific to a particular type of annotation, and they offered limited possibilities to combine information from different annotation layers applied to the same piece of text. Such multi-layer corpora became increasingly popular,[1] and, more importantly, they represent a valuable source to study interdependencies between different types of annotation. For example, the development of a semantic parser usually takes a syntactic analysis as its input, and higher levels of linguistic analysis, e.g., coreference resolution or discourse structure, may take both types of information into consideration. Such studies, however, require that all types of annotation applied to a particular document are integrated into a common representation that provides lossless and comfortable access to the linguistic information conveyed in the annotation without requiring too laborious conversion steps in advance.

This has been one motivation for research on interoperability between different types of annotation. Another motivation was that different NLP tools for the same linguistic phenomenon, say, syntactic parsers, produce different output formats, and that comparative evaluations as well as ensemble combination architectures require a interoperable representation of the respective analyses (Pareja-Lora and Aguado de Cea, 2010; Chiarcos, 2010b). And with the development of complex NLP pipeline systems, people became interested in interchangeable pipeline modules, where one module can be replaced by another, equivalent module, for example, if domain-specific parsers are to be used if texts from their particular domain are to be analyzed. This would be possible, however, only if these modules make use of the same input and output representations, and if they refer to a common vocabulary of linguistic categories (Buyko et al., 2008).

In this paper, I focus on **interoperability of linguistic corpora**. This is closely related to interoperability in NLP pipelines (cf. Hellmann et al., this vol., p. 185ff.), but differs in two crucial aspects: (1) In an NLP pipelines, annotations can be created on-the-fly, and do not necessarily have to be preserved throughout the entire pipeline. It is therefore not necessary to formally distinguish different layers of an-

---

[1] For example, parts of the Penn Treebank (Marcus et al., 1993), originally annotated for parts-of-speech and syntax, were later annotated with nominal semantics, semantic roles, time and event semantics, discourse structure and anaphoric coreference (Pustejovsky et al., 2005).

notation, to represent the macrostructure of linguistic corpora and metadata. (2) NLP pipelines are usually created for one particular task or a set of tasks. The number of possible annotation layers is thus limited by plausibility considerations.[2] For linguistic corpora, however, the number and the types of annotations applied to a particular text are in principle unlimited, because researchers may have an interest to preserve and to integrate all available legacy annotations created throughout the entire lifetime of a particular corpus. In that sense, corpus interoperability is a more general and harder problem than NLP interoperability.

At the moment, state-of-the-art approaches on **structural interoperability** of linguistic corpora build on standoff-XML (Carletta et al., 2005; Ide and Suderman, 2007; Bouda and Cysouw, this vol.) and relational data bases (Chiarcos et al., 2008; Eckart et al., this vol.). The underlying data models are, however, graph-based, and Sect. 2 pursues the idea that RDF and RDF data bases can be applied for the task to represent all possible annotations of a corpus in an interoperable way, to integrate their information without any restrictions (as imposed, for example, by conflicting hierarchies or overlapping segments in an XML-based format), and to provide means to store and to query this information regardless of the annotation layer from which it originates. Using OWL/DL defined data types as the basis of this RDF representation allows to specify and to verify formal constraints on the correct representation of linguistic corpora in RDF. POWLA, the approach presented here, formalizes data models for generic linguistic data structures for linguistic corpora as OWL/DL concepts and definitions (POWLA TBox) and represents the data as OWL/DL individuals in RDF (POWLA ABox).

The heterogeneity of linguistic annotations has long been recognized as a key problem limiting the reusability of NLP tools and linguistic data collections, and it is generally agreed that repositories of linguistic annotation terminology represent a key element in the establishment of **conceptual interoperability**. With a terminological reference repository, it is possible to overcome the heterogeneity of annotation schemes: Reference definitions provide an interlingua that allows mapping linguistic annotations from annotation scheme A to annotations in accordance with scheme B. Several repositories of linguistic annotation terminology have been developed by the NLP/computational linguistics community (Leech and Wilson, 1996; Aguado de Cea et al., 2004; Pareja-Lora, this vol.) as well as in the field of language documentation/typology (Bickel and Nichols, 2002; Saulwick et al., 2005), and their continuous application is expected to enhance the consistency of linguistic metadata and annotations. The General Ontology of Linguistic Description (Farrar and Langendoen, 2003, GOLD) and the ISO TC37/SC4 Data Category Registry (Kemps-Snijders et al., 2009; Windhouwer and Wright, this vol., ISOcat) address both communities.

At the moment, however, a problems for the practical application of these terminology repositories persists with the fact that different communities develop and maintain terminology repositories – e.g., GOLD and ISOcat – independently, and

---

[2] In the general case, for example, one would expect that there is only a single syntactic analysis performed, but not several parses whose input is integrated.

these repositories are not always compatible with respect to the definitions they provide,[3] with respect to the technologies employed,[4] or with respect to the underlying philosophy.[5] Researchers are aware of the problem and actively addressing it, e.g., by providing ISOcat in OWL (like GOLD, cf. Windhouwer and Wright, this vol.) or by integrating GOLD categories as a separate profile in ISOcat (Kemps-Snijders, 2010). Section 3 describes the Ontologies of Linguistic Annotation (OLiA ontologies) that introduce an intermediate level of representation between ISOcat, GOLD and other repositories of linguistic reference terminology in order to facilitate the development of applications and resources that take benefit of a well-defined terminological backbone even **before** the GOLD and ISOcat repositories have converged into a uniform and generally accepted reference terminology.

A novel element in this approach is that conceptual interoperability and structural interoperability are addressed with the same formalism (OWL/RDF). In earlier approaches, e.g., Chiarcos et al. (2008), these were treated independently, and conceptual interoperability was established through a software-mediated mapping between annotations and ontologies. This paper describes a fully declarative approach.

Moreover, by using RDF as representation format, both resources described here, POWLA corpora and OLiA ontologies, can be integrated with resources already available as Linked Data, e.g., meta data repositories such as Glottolog/Langdoc (Nordhoff, this vol.), general-purpose knowledge bases like the DBpedia (Hellmann et al., this vol.) or full-fledged lexical-semantic resources such as the Wikipedia and WordNet (McCrae et al., this vol.).

2 Addressing Structural Interoperability

POWLA is an OWL/DL-based formalism to represent linguistic corpora in an interoperable way. As compared to earlier approaches in this direction (Burchardt et al., 2008; Hellmann et al., 2010), POWLA is not tied to a specific selection of annotation layers, or a specific annotation scheme. Instead, it is designed to support any kind of text-oriented annotation.

---

[3] As one example, the original GOLD Numeral was a Determiner (Numeral ⊑ Quantifier ⊑ Determiner, http://linguistics-ontology.org/gold/2009/Numeral), whereas a ISOcat Numeral (DC-1334) is defined on the basis of its semantic function, without any references to syntactic categories. Thus, *two* in *two of them* may be a ISOcat Numeral but not a GOLD Numeral. Following a suggestion of the author, the current GOLD version (http://linguistics-ontology.org/gold/Numeral) adopted the ISOcat modeling, but as GOLD and ISOcat implement community processes on different communities, the establishment of parallel definitions takes a considerable time, if it is possible at all.

[4] GOLD is based on OWL/RDF, a formalization in OWL/DL is possible, it thereby supports the full power of description logics (i.e. a decidable fragment of first-order predicate logic). ISOcat, however, is designed as a semistructured list of concepts, with only optional elements of hierarchical organization.

[5] GOLD aims to provide a holistic and unified representation of linguistic reference concepts. ISOcat is an extensible collection of annotation categories.

The idea underlying POWLA is to represent linguistic annotations by means of RDF, to employ OWL/DL to define data types and consistency constraints for these RDF data, and to adopt these data types and constraints from an existing representation formalism applied for the loss-less representation of arbitrary kinds of text-oriented linguistic annotation within a generic exchange format. POWLA is designed as an OWL/DL implementation of the PAULA data model (Dipper, 2005; Chiarcos et al., 2008, 2011) developed at the Collaborative Research Center (SFB) 632 "Information Structure" at the University of Potsdam, Germany. At the moment, the standard linearization of PAULA is an XML standoff format that originates from early drafts of the Linguistic Annotation Framework (Ide and Romary, 2004), and it is thus closely related to the later ISO TC37/SC4 format GrAF (Ide and Suderman, 2007). With POWLA as an OWL/DL linearization of the PAULA data model, all annotations currently covered by PAULA (presumably every kind of text-oriented linguistic annotation) can be represented as part of the Linguistic Linked Open Data cloud.

2.1 PAULA Data Types

PAULA implements the insight that any kind of linguistic annotation can be represented by means of **directed (acyclic) graphs** (Bird and Liberman, 2001; Ide and Suderman, 2007), i.e. the basic triple structure underlying RDF: Aside from the primary data (text), linguistic annotations consist of three principal components, i.e. segments (spans of text, e.g. a phrase, modeled as nodes), relations between segments (e.g. dominance relation between two phrases, modeled as edges) and annotations that describe different types of segments or relations (modeled as labels).

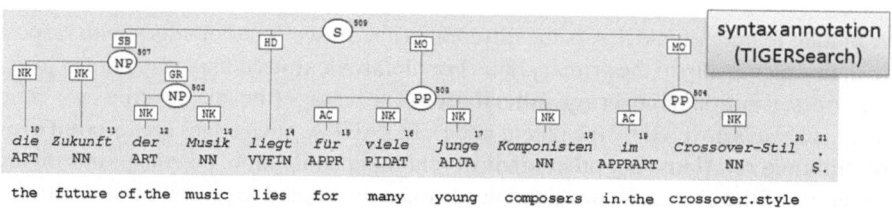

Fig. 1 Constituent syntax: Example from the NEGRA corpus (Skut et al., 1998), visualization based on TIGERSearch (Lezius, 2002).

PAULA data types relevant for linguistic annotations are the following (see Fig. 2 for the German phrase *für viele junge Komponisten* 'for many young composers' as an illustration, for the original syntax visualization see Fig. 1):

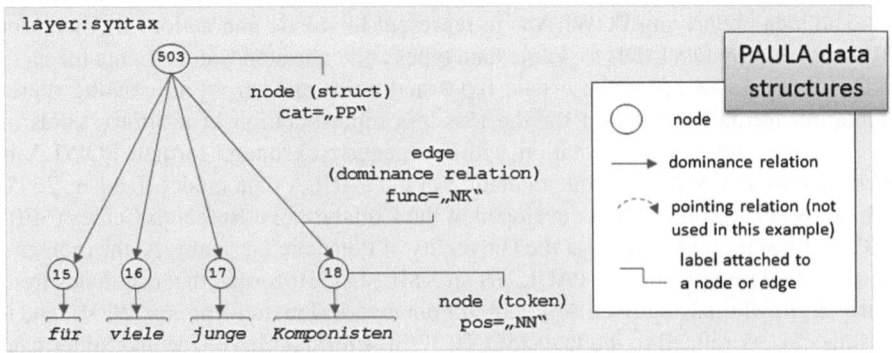

Fig. 2 Constituent syntax: Fragment from Fig. 1 with PAULA data structures.

node (structural units of annotation)
  token   character spans in the primary data
  markable   span of tokens (data structure of flat, layer-based annota-
                             tions defined with respect to, e.g. a timeline)
  struct   hierarchical data structure forming DAGs (e.g. trees) by
                             establishing parent-child relations between a struct (pa-
                             rent) and tokens, markables or other structs.
edge (relational unit of annotation, connecting nodes)
  dominance relation directed edge between a struct and its children, coverage
                             inheritance (see below)
  pointing relation   general directed edge, no coverage inheritance
label (attached to nodes or edges)
  features   linguistic annotations

A unique feature of PAULA is the differentiation of two types of edges with respect to their relationship to the primary data. For hierarchical structures, e.g. phrase structure trees, a notion of **coverage inheritance** is necessary (the text covered by a child node is always covered by the parent node). In PAULA, such edges are referred to as **dominance relations**. For other kinds of relational annotation, no constraints on the coverage of the elements connected needs to be postulated (e.g. anaphoric relations, alignment in parallel corpora, dependency analyses), and source and target of a relation may or may not overlap at all. In PAULA, edges without coverage inheritance are referred to as **pointing relations**. This distinction does not constrain the generic character of PAULA (a general directed graph would just use pointing relations), but it captures a fundamental distinction of linguistic data types. As such, it was essential for the development of convenient means of visualization and querying of PAULA data: For example, the appropriate visualization (hierarchical or relational) within a corpus management system can be chosen on the basis of the data structures alone, and it does not require any external specifications (Chiarcos et al., 2008).

Additionally, PAULA includes specifications for the organization of annotations, which is, however, skipped here for reasons of space.

2.2 The POWLA Ontology (POWLA TBox)

The POWLA ontology represents a straight-forward implementation of the PAULA data types in OWL/DL. Document, Relation; Node and Layer are defined as subclasses of POWLAElement. Here, we concentrate on Node and Relation, Document and Layer are more important for corpus organization.

A POWLAElement is anything that can carry a label (property hasLabel). For Node and Relation, hasLabel contains string values of linguistic annotation (subproperty hasAnnotation). The property hasAnnotation are, however, not to be used directly, but rather, subproperties are to be created for every annotation phenomenon, e.g. has_pos for part-of-speech annotation.

A Node is a POWLAElement that covers a stretch of primary data. It can carry hasChild properties (and the inverse hasParent) that express coverage inheritance. A Relation is another POWLAElement that is used for every edge that carries an annotation. The properties hasSource and hasTarget (resp. the inverse isSourceOf and isTargetOf) assign a Relation source and target node. Dominance relations are relations whose source and target are connected by hasChild, pointing relations are relations where source and target are not connected by hasChild. It is thus not necessary to distinguish pointing relations and dominance relations as separate concepts in the POWLA ontology.

Two basic subclasses of Node are distinguished: A Terminal is a Node that does not have a hasChild property. It corresponds to a "token" in PAULA. A Nonterminal is a Node that has at least one hasChild property. The differentiation between PAULA struct and markable can be inferred and is therefore not explicitly represented in the ontology: A struct is Nonterminal that has another Nonterminal as its child, or that is connected to at least one of its children by means of a (dominance) Relation, any other Nonterminal corresponds to a PAULA markable.

The concept Root was introduced for organizational reasons. It corresponds to a Nonterminal that does not have a parent, i.e. the top-level node within an annotation layer (and may be either a Terminal or a Nonterminal).

Both Terminals and Nonterminals are characterized by a string value (property hasString), and a particular position (properties hasStart and hasEnd) with respect to the primary data. Terminals are further connected with each other by means of nextTerminal properties. This is, however, a preliminary solution. Forthcoming versions of POWLA may address Nonterminals by means of pre- and post-order as defined by Trißl and Leser (2007), and Terminals may be linked to strings in accordance to the NLP Interchange Format NIF (cf. Hellmann et al., this vol., p. 185ff.).

2.3 Modelling Linguistic Annotations in POWLA (POWLA ABox)

The POWLA ontology defines data types that can now be used to represent linguistic annotations. Figure 3 shows the `Nonterminal` created for the phrase *für viele junge Komponisten* 'for many young composers', the `Terminal` for *Komponisten* 'composers', and the `Relation` between them.

```
<powla:Nonterminal rdf:about="s1_503">
    <powla:hasLayer rdf:resource="syntax"/>
    <powla:has_cat>PP</powla:has_cat>
    <powla:hasChild rdf:resource="s1_18"/>
    ...
</powla:Nonterminal>

<powla:Relation rdf:about="s1_503_to_s1_18">
    <powla:hasLayer rdf:resource="syntax"/>
    <powla:has_func>NK</powla:has_func>
    <powla:hasSource rdf:resource="s1_503"/>
    <powla:hasTarget rdf:resource="s1_18"/>
</powla:Relation>

<powla:Terminal rdf:about="s1_18">
    <powla:hasLayer rdf:resource="syntax"/>
    <powla:hasString>Komponisten</powla:hasString>
    <powla:has_pos>NN</powla:has_pos>
    <powla:nextTerminal rdf:resource="s1_19"/>
    <powla:startPosition>103</powla:startPosition>
    <powla:endPosition>113</powla:endPosition>
</powla:Terminal>
```

Fig. 3 Constituent syntax: A `Nonterminal`, a `Terminal`, and the `Relation` between them in POWLA, fragment of Fig. 2.

The `Nodes` are taken from the German sentence analyzed in Fig. 1. The `Node` IDs preserve the original identifiers used in the NEGRA corpus (Skut et al., 1998), with `s1_18` corresponding to the 18th word, and `s1_503` corresponding to the phrase with ID 503 in sentence 1.[6] The `Relation` ID is derived from the IDs of the source and the target node. The properties `has_pos`, `has_cat` and `has_func` are subproperties of `hasAnnotation` that have been created to reflect the `pos`, `cat` and `func` labels of nodes and edges in Fig. 1. `Relation s1_503_to_s1_18` is marked as a dominance relation by the accompanying `hasChild` relation between its source and target.

It should be noted that the RDF representation given in Fig. 3 is by no means complete. Inverse properties, for example, are missing, e.g., the `hasParent` rela-

---

[6] The data was not directly converted from the NEGRA format, but through TIGER XML. These naming conventions, as well as those of the attribute names (`func` instead of `edge`, `pos` instead of `tag`) originate from the converter integrated in the TIGERRegistry (Lezius, 2002).

tion between s1_18 and s1_503. Using a reasoner, however, the missing RDF triples can be inferred from the information provided explicitly. This inference mechanism can be also be applied to infer ⊑ (rdfs:subClassOf) relationships (e.g., that s1_18 is not only a Terminal, but also a Node), and the transitive closure of relations (if the corresponding transitivity axioms are added). Further, functional differentiations can be inferred, e.g., the difference between PAULA markables and structs. This is not relevant for processing, but for visualization only, and therefore explicitly represented in PAULA XML. In POWLA, it can be inferred whether a Layer requires a spreadsheet-like visualization ('grid view', applicable to PAULA markable layers, i.e., a POWLA Layer with no recursive structures and no labeled dominance relations), or as a directed acyclic graph (e.g., a tree, applicable to PAULA struct layers, i.e., a POWLA Layer with recursion or labeled dominance relations). Moreover, a reasoner would also allow us to verify whether the necessary cardinality constraints are respected, e.g., every Relation has exactly one hasSource and one hasTarget relation etc.

Although illustrated here for syntax annotations only, other annotation layers represented in PAULA XML can be equally easily transformed to POWLA. Using existing converters that generate PAULA XML, e.g., Salt'N'Pepper (Zipser and Romary, 2010), a broad variety of input formats from various tools are supported, including TIGER XML (König and Lezius, 2000, constituent and dependency syntax), MMAX2 (Müller and Strube, 2006, coreference annotation), EXMARaLDA (Schmidt, 2004, transcriptions and layer-based annotation), RSTTool (O'Donnell, 2000, discourse structure annotation), and Toolbox (Busemann and Busemann, 2008, typological glosses).[7]

Like PAULA XML, POWLA represents these different kinds of annotations in an interoperable way, but here on the basis of OWL and RDF. One important difference between PAULA XML and POWLA is that POWLA data can be directly fed into an RDF triple store, whereas current data base solutions for PAULA involve relational data bases and the transformation into a proprietary table format (Zeldes et al., 2009). Another important difference is that all IDs used in the POWLA representation are URIs, with an XML base name as specified in the file they are contained in. Accordingly, these URIs can be referred to from external resources, e.g., from scientific papers (as stable references to corpus examples) or from lexical-semantic resource (as examples to illustrate a specific semantic role). And of course, POWLA corpora can be augmented with references to other resources available as Linked Data, e.g., terminology repositories as described in the following section, or meta data repositories as, for example, described by Nordhoff (this vol.).

---

[7] Through existing converters that produce these formats, numerous additional formats are supported, e.g., Penn Treebank syntax annotations (Marcus et al., 1994, through TIGER XML), Praat (Boersma, 2002, through EXMARaLDA), ELAN (Hellwig et al., 2008, through EXMARaLDA), etc. Additionally, PAULA XML wrappers for a number of NLP tools exist, which cover part-of-speech tagging, parsing, anaphor resolution, connective classification and discourse parsing, as applied, for example, in the text summarization pipeline described by Stede et al. (2006).

3 Addressing Conceptual Interoperability

The Ontologies of Linguistic Annotation (OLiA) are a repository of annotation terminology for various linguistic phenomena on a great band-width of languages. In combination with RDF-based formats like POWLA (Sect. 2) and NIF (Hellmann et al., this vol., Sect. 4), the OLiA ontologies allow to represent linguistic annotations and their meaning within the Linguistic Linked Open Data cloud in an interoperable way.

3.1 Towards Conceptual Interoperability of Linguistic Annotations

The OLiA ontologies introduce an intermediate level of representation between ISOcat, GOLD and other repositories of linguistic reference terminology and are interconnected with these resources, and they provide not only a means to formalize reference categories, but also annotation schemes, and the way that these are linked with reference categories.

3.2 A Modular Architecture of OWL/DL Ontologies

The Ontologies of Linguistic Annotations – briefly, OLiA ontologies (Chiarcos, 2008) – represent a modular architecture of OWL/DL ontologies that formalize several intermediate steps of the mapping between annotations, a 'Reference Model' and existing terminology repositories ('External Reference Models').

The OLiA ontologies were developed as part of an infrastructure for the sustainable maintenance of linguistic resources (Schmidt et al., 2006), and their primary fields of application include the formalization of annotation schemes and concept-based querying over heterogeneously annotated corpora (Rehm et al., 2007; Chiarcos et al., 2008).

In the OLiA architecture, four different types of ontologies are distinguished:

- The OLIA REFERENCE MODEL specifies the common terminology that different annotation schemes can refer to. It is derived from existing repositories of annotation terminology and extended in accordance with the annotation schemes that it was applied to.
- Multiple OLIA ANNOTATION MODELs formalize annotation schemes and tagsets. Annotation Models are based on the original documentation of an annotation scheme, they provide an interpretation-independent representation.
- For every Annotation Model, a LINKING MODEL defines rdfs:subClassOf (⊑) relationships between concepts/properties in the respective Annotation Model and the Reference Model. Linking Models are interpretations of Annotation Model concepts and properties in terms of the Reference Model.

- Existing terminology repositories can be integrated as EXTERNAL REFERENCE MODELs, if they are represented in OWL/DL. Then, Linking Models specify ⊑ relationships between Reference Model concepts and External Reference Model concepts.

The OLiA Reference Model specifies classes for linguistic categories (e.g. olia:Determiner) and grammatical features (e.g. olia:Accusative), as well as properties that define relations between these (e.g. olia:hasCase). Far from being yet another annotation terminology ontology, the OLiA Reference Model does not introduce its own view on the linguistic world, but rather, it is a derivative of the EAGLES recommendations (Leech and Wilson, 1996), MUL-TEXT/East (Erjavec, 2004), and GOLD (Farrar and Langendoen, 2003) that was introduced as a technical means to allow to interpret linguistic annotations with respect to these terminological repositories and extended with respect to the annotation schemes linked with it. These extensions are also further communicated to the communities behind GOLD and ISOcat. The Reference Model specifies for example that olia:PrepositionalPhrase ⊑ olia:NounHeadedPhrase.[8]

Annotation Models differ conceptually from the Reference Model in that they include not only concepts and properties, but also individuals: Individuals represent concrete tags, while classes represent abstract concepts similar to those of the Reference Model. As an example, consider the tag PP from the syntax annotation the NEGRA corpus Skut et al. (1997) and the corresponding individual tiger:prepositionalPhrase in the Annotation Model http://purl.org/olia/tiger-syntax.owl:[9]

    tiger:prepositionalPhrase system:hasTag 'PP'
  tiger:prepositionalPhrase a tiger:PrepositionalPhrase
Linking Models then import an Annotation Model and the Reference Model and specify relations between their concepts: tiger:PrepositionalPhrase ⊑ olia:PrepositionalPhrase. The Linking with External Reference Models like ISOcat is analogous: olia:PrepositionalPhrase ⊑ isocat:DC-2257. In consequence, it is true that tiger:prepositionalPhrase rdf:type isocat:DC-2257

---

[8] olia:NounHeadedPhrase was introduced as a generalization over prepositional phrase and noun phrase to account for the constituent representation of certain dependency parsers where this differentiation is not made. However, there is little agreement whether the noun or the preposition is the head of a prepositional phrase, but this debate is independent of the modeling of the OLiA Reference Model: The *linking* with external reference models has to specify that the olia:NounHeadedPhrase corresponds to a single, well-defined category in an independently developed community-maintained terminology repository, or that it does not (then, a disjunction would be necessary).

[9] Because the annotation scheme of the NEGRA corpus is closely related with the annotation scheme of the TIGER corpus (Brants et al., 2004), both are represented together in the Annotation Model developed for TIGER-style syntax.

Within an application, the German phrase *für viele junge Komponisten* considered above can then be circumscribed by means of the concepts it is associated with:[10]

```
olia:PrepositionalPhrase and
powla:hasChild some (olia:CommonNoun and
                     olia:hasGender some olia:Masculine and
                     olia:hasNumber some olia:Plural) and
    ...
```

This description is concept-based and thus independent from any particular tagset, and applied to another Annotation Model, a query for `olia:PrepositionalPhrase` would retrieve another set of individuals that represent the same meaning with different annotations, e.g., the phrase *von den geschäftlichen Revisoren* 'by the business auditors' from the TüBa-D/Z corpus (document 1, sentence 19)[11] – even though it carries the label PX and not PP like in the NEGRA/TIGER annotation scheme.[12]

3.3 Current Status of the OLiA Ontologies

The OLiA ontologies are available from `http://purl.org/olia`. At the moment, they have not been officially released, they will be released under a Creative Commons Attribution license in mid-2012.

The OLiA ontologies cover different grammatical phenomena, including inflectional morphology, word classes, phrase and edge labels of different syntax annotations, as well as discourse annotations (coreference, discourse relations, discourse structure and information structure). Annotations for lexical semantics are only covered by the OLiA ontologies to the extent that they are encoded in syntactic and morphosyntactic annotation schemes (e.g. as grammatical roles). For lexical semantic annotations in general, a number of reference resources are already available as Linked Data, including RDF versions of WordNet,[13] FrameNet,[14] and the Wikipedia.[15]

The OLiA Reference Model comprises 14 `MorphologicalCategorys` (morphemes), 263 `MorphosyntacticCategorys` (word classes/part-of-speech

---

[10] The linking between corpus data and the OLiA ontologies can be accomplished, for example, by copying all properties of an OLiA Annotation Model individual to the POWLA individuals with the corresponding annotations, as specified by the `hasTag` property.

[11] `http://www.sfs.uni-tuebingen.de/tuebadz.shtml`

[12] See `http://purl.org/olia/tueba.owl#PX` and `http://purl.org/olia/tueba-link.rdf`.

[13] `http://thedatahub.org/dataset/w3c-wordnet`, also see McCrae et al. (this vol.).

[14] `http://www.loa.istc.cnr.it/codeps/owl/ofntb.owl`, cf. Nuzzolese et al. (2011).

[15] `http://dbpedia.org`, cf. Hellmann et al. (this vol.).

tags), 83 `SyntacticCategorys` (phrase labels), and 326 different values for 16 `MorphosyntacticFeatures`, 4 `MorphologicalFeatures`, 4 `SyntacticFeatures` and 4 `SemanticFeatures`.

As for morphological, morphosyntactic and syntactic annotations, the OLiA ontologies include 32 Annotation Models for about 70 different languages, including several multi-lingual annotation schemes, e.g. the EAGLES recommendations (Chiarcos, 2008) for 11 Western European languages, and Multext-East (Chiarcos and Erjavec, 2011) for 15 (mostly) Eastern European languages. As for non-(Indo-)European languages, the OLiA ontologies include morphosyntactic annotation schemes for languages of the Indian subcontinent, for Arabic, Basque, Chinese, Estonian, Finnish, Hausa, Hungarian and Turkish. Other languages, including languages of Africa, the Americas, the Pacific and Australia are covered by Annotation Models developed for glosses as produced in typology and language documentation. The OLiA ontologies also cover historical language stages, including Old High German, Old Norse and Old/Classical Tibetan.

As mentioned above, application of modular OWL/DL ontologies allows to link annotations with terminological repositories: Annotation schemes and reference terminology are formalized as OWL/DL ontologies, and the linking is specified by `rdfs:subClassOf` descriptions. This mechanism has also been applied to link the OLiA Reference Model with existing terminology repositories, including GOLD (Chiarcos, 2008), the OntoTag ontologies (Buyko et al., 2008, cf. Pareja-Lora, this vol.) and ISOcat (Chiarcos, 2010a, cf. Windhouwer and Wright, this vol.). Thereby, the OLiA Reference Model provides a stable intermediate representation between existing terminology repositories and ontological models of annotation schemes. This allows any concept that can be expressed in terms of the OLiA Reference Model also to be interpreted in the context of ISOcat, GOLD or the OntoTag ontologies. Using the OLiA Reference Model, it is thus possible to develop applications that are interoperable in terms of GOLD *and* ISOcat even though both are still under development and both differ in their conceptualizations (see footnote 3).

Example applications of the OLiA ontologies include the specification of grammatical features in lexical resources (McCrae et al., 2011, this vol.), the development of tagset independent NLP architectures (Chiarcos, 2010b; Hellmann et al., this vol.), and tagset independent corpus queries, e.g. the combination of OLiA reference concepts with SPARQL queries on POWLA data, as shown in the following section.

4 A Use-Case for Corpus Interoperability: Developing Resource-Independent Corpus Queries

A number of different applications of OWL/RDF-encoded corpora are possible. From the perspective of corpus linguistics and NLP, it is essential that the data structure provides exhaustive access to the information conveyed in the corpus. This can be shown, for example, by implementing a **corpus querying engine**. A pilot experiment has been performed where a multi-layer corpus, the syntax-annotated German NEGRA corpus (Skut et al., 1998) with the coreference annotations by Schiehlen (2004) was converted to POWLA. The RDF data was loaded in an RDF database, OpenLink Virtuoso, and could thus be queried with SPARQL. To illustrate that SPARQL queries can retrieve all relevant information from POWLA data, the operators of the ANNIS query language (AQL) were reimplemented as SPARQL macros, the query language used in the corpus querying system ANNIS (Chiarcos et al., 2008). With ANNIS, a query engine for PAULA data is available, and was successfully applied in a number of linguistic and philological projects, so, it can be assumed that AQL represents the minimal power necessary to explore any kind of linguistic corpora. The portability of PAULA to the Semantic Web could thus be shown with the reconstruction of all AQL operators as SPARQL macros.[16] Although detailed results of this evaluation are described elsewhere, we can state here that multi-layer corpora in POWLA can be queried with SPARQL macros in basically the same way as with AQL.

SPARQL actually provides us with an even more powerful means of querying: For example, AQL does not support queries for the absence of a particular annotation (e.g. an NP not dominating a pronoun), which can be easily expressed in SPARQL:

```
PREFIX negra: <http://purl.org/powla/negra-sample.owl#>.
PREFIX powla: <http://purl.org/powla/powla.owl#>.
SELECT DISTINCT ?np
WHERE {
    ?np a powla:Nonterminal.
    ?np negra:has_cat "NP".
    OPTIONAL { ?np powla:hasChild ?pronoun.
               ?pronoun negra:has_pos "PPER" }
    FILTER !bound(?pronoun)
}
```

POWLA can thus be used to create linguistic information systems. An additional advantage of the OWL/RDF formalization is that it represents a standardized to represent heterogeneous data collections: With RDF, a standardized representation formalism for different corpora is available, and with data types being defined in OWL/DL, the validity of corpora can be automatically checked (according to the consistency constraints posited by the POWLA ontology). POWLA represents a possible solution to the **structural interoperability** challenge for linguistic corpora (Ide and Pustejovsky, 2010). Unlike other formalisms developed in this direction (e.g., PAULA, or GrAF, Ide and Suderman, 2007), it does not involve a special-purpose XML standoff format, but it builds on established standards with broad technical support and an active and comparably large community. Standard formats specifically designed for linguistic annotations as developed in the context of ISO TC37/SC4 (e.g. GrAF, Ide and Suderman, 2007 and TIGER2, Romary et al., 2011), are, however, still under development.

---

[16] Details for this conversion, a sample data and SPARQL macros for querying this data can be found under http://purl.org/powla.

In comparison of this approach with current initiatives within the linguistics/NLP community, e.g. ISO TC37/SC4, which focus on complex standoff XML formats specifically designed for linguistic data, this approach offers three crucial advantages:

1. The increasing number of RDF data bases provides us with convenient means for the management of linguistic data collections.
2. Augmenting an RDF representation of linguistic corpora with formally specified consistency conditions, i.e., an OWL/DL specification of data types and constraints, existing reasoners can be applied to check the consistency of this representation.
3. Resources can be freely interconnected with each other and with lexical-semantic resources available from the Linked Open Data cloud.

Additionally, an RDF representation of linguistic corpora allows for their **integration with other RDF resources**. The linking of POWLA corpora with OLiA ontologies as described before, for example, allows reformulating the SPARQL query for NPs not dominating a pronoun, such that the reformulated query can also be applied to corpora with other annotation schemes and is thus, interoperable:

```
WHERE {
    ?np a olia:NounPhrase.
    OPTIONAL { ?np powla:hasChild ?pronoun.
               ?pronoun a olia:PersonalPronoun }
    FILTER !bound(?pronoun)
}
```

Representing metadata and annotations of a corpus by means of references to resources in the LOD cloud, it becomes possible to define metadata filters for linguistic corpora and queries for linguistic annotations that are independent of the string representation of this information in the corpora. Resources represented in this way are thus not only structurally, but also **conceptually interoperable**.

If published as Linked Data, corpora represented in RDF can further be **linked with lexical-semantic resources** already available as Linked Data, e.g., a general knowledge base like DBpedia (Hellmann et al., this vol.), or linguistic resources like WordNet or Wiktionary (McCrae et al., this vol.). Semantic annotations, e.g., those of PropBank (Kingsbury and Palmer, 2002), can therefore be implemented as a linking between corpora and semantic resources, without the need to duplicate or to synchronize resources obtained from different providers.

Another important type of resources available as Linked Data includes **repositories of metadata and terminology**. Lexvo,[17] for example, provides identifiers for languages based on ISO 639; Glottolog (Nordhoff, this vol.) provides an even more fine-grained taxonomy of languoids.

A unique feature of the approach described here is that OWL and RDF provide a solution for both aspects of corpus interoperability – and even interoperability between corpora and lexical-semantic resources – **within the same formalism**. One

---

[17] http://lexvo.org

concrete advantage of such a holistic approach is that novel corpus information systems can be developed that achieve interoperability by simpler means, e.g., using a single data base for both corpus data and annotation mapping, whereas traditional systems that combine XML- or SQL-based corpus representations with linguistic ontologies (e.g., Rehm et al., 2007; Chiarcos et al., 2008) require a more complex system architecture and thus, greater implementation efforts.[18]

Acknowledgements The research described here was partially supported by the German Research Foundation (DFG) through the Collaborative Research Center (SFB) 441 "Linguistic Data Structures" (University of Tübingen, 2006-2008, OLiA ontologies) and through the Collaborative Research Center (SFB) 632 "Information Structure" (University of Potsdam, 2008-2011, OLiA ontologies, POWLA)

References

Bickel B, Nichols J (2002) Autotypologizing databases and their use in fieldwork. In: Proceedings of the LREC-2002 Workshop on Resources and Tools in Field Linguistics, Las Palmas, Spain

Bird S, Liberman M (2001) A formal framework for linguistic annotation. Speech Communication 33(1-2):23–60

Boersma P (2002) Praat, a system for doing phonetics by computer. Glot international 5(9/10):341–345

Bouda P, Cysouw M (this vol.) Treating dictionaries as a Linked-Data corpus. pp 15–23

Brants S, Dipper S, Eisenberg P, Hansen S, König E, Lezius W, Rohrer C, Smith G, Uszkoreit H (2004) TIGER: Linguistic interpretation of a German corpus. Research on Language and Computation 2(4):597–620

Burchardt A, Padó S, Spohr D, Frank A, Heid U (2008) Formalising Multi-layer Corpora in OWL/DL – Lexicon Modelling, Querying and Consistency Control. In: Proceedings of the 3rd International Joint Conference on NLP (IJCNLP 2008), Hyderabad

Busemann A, Busemann K (2008) Toolbox self-training. Tech. rep., http://www.sil.org. Version 1.5.4, Oct 2008

Buyko E, Chiarcos C, Pareja-Lora A (2008) Ontology-based interface specifications for a NLP pipeline architecture. In: Proceedings of the 6th International Conference on Language Resources and Evaluation (LREC 2008), Marrakech, Morocco

---

[18] Another crucial advantage of RDF data bases as compared to relational data bases is that RDF data can be flexibly added or removed. For debugging purposes, the OWL/DL-defined data model (and the RDF triples inferred) can thus be adjusted without without reinitializing the data base, thereby substantially accelerating development cycles.
As compared to XML data bases, RDF data dases support multi-layer corpora with an unrestricted band-width of annotations, whereas XML data bases are optimized for hierarchical annotations (e.g., syntax trees without crossing edges and overlapping segments), but relatively inefficient for non-hierarchical data (e.g., coreference, overlapping hierarchies in multi-layer corpora).

Carletta J, Evert S, Heid U, Kilgour J (2005) The NITE XML Toolkit: data model and query. Language Resources and Evaluation Journal (LREJ) 39(4):313–334

Aguado de Cea G, Gomez-Perez A, Alvarez de Mon I, Pareja-Lora A (2004) Onto-Tag's linguistic ontologies. In: Proc. Information Technology: Coding and Computing (ITCC'04), Washington, DC, USA

Chiarcos C (2008) An ontology of linguistic annotations. LDV Forum 23(1):1–16

Chiarcos C (2010a) Grounding an ontology of linguistic annotations in the Data Category Registry. In: LREC 2010 Workshop on Language Resource and Language Technology Standards (LT&LTS), Valetta, Malta, pp 37–40

Chiarcos C (2010b) Towards robust multi-tool tagging. An OWL/DL-based approach. In: ACL 2010, Uppsala, Sweden, pp 659–670

Chiarcos C, Erjavec T (2011) OWL/DL formalization of the MULTEXT-East morphosyntactic specifications. In: 5th Linguistic Annotation Workshop, Portland, pp 11–20

Chiarcos C, Dipper S, Götze M, Leser U, Lüdeling A, Ritz J, Stede M (2008) A Flexible Framework for Integrating Annotations from Different Tools and Tagsets. TAL (Traitement automatique des langues) 49(2)

Chiarcos C, Ritz J, Stede M (2011) By all these lovely tokens ... Merging conflicting tokenizations. Journal of Language Resources and Evaluation (LREJ) 4(45). to appear

Dipper S (2005) XML-based stand-off representation and exploitation of multi-level linguistic annotation. In: Proc. Berliner XML Tage 2005 (BXML 2005), Berlin, Germany, pp 39–50

Eckart K, Riester A, Schweitzer K (this vol.) A discourse information radio news database for linguistic analysis. pp 65–75

Erjavec T (2004) MULTEXT-East version 3: Multilingual morphosyntactic specifications, lexicons and corpora. In: Fourth International Conference on Language Resources and Evaluation, (LREC 2004), Lisboa, Portugal, pp 1535–1538

Farrar S, Langendoen D (2003) A linguistic ontology for the semantic web. Glot International 7(3):97–100

Hellmann S, Unbehauen J, Chiarcos C, Ngonga Ngomo A (2010) The TIGER Corpus Navigator. In: 9th International Workshop on Treebanks and Linguistic Theories (TLT-9), Tartu, Estonia, pp 91–102

Hellmann S, Stadler C, Lehmann J (this vol.) The German DBpedia: A sense repository for linking entities. pp 181–189

Hellwig B, Uytvanck DV, Hulsbosch M (2008) ELAN - Linguistic Annotator. Tech. rep., http://www.lat-mpi.eu/tools/elan. version of 2008-07-31

Ide N, Pustejovsky J (2010) What does interoperability mean, anyway? Toward an operational definition of interoperability. In: Proc. Second International Conference on Global Interoperability for Language Resources (ICGL 2010), Hong Kong, China

Ide N, Romary L (2004) International standard for a linguistic annotation framework. Natural language engineering 10(3-4):211–225

Ide N, Suderman K (2007) GrAF: A graph-based format for linguistic annotations. In: Proc. Linguistic Annotation Workshop (LAW 2007), Prague, Czech Republic, pp 1–8

Kemps-Snijders M (2010) Relish: Rendering endangered languages lexicons interoperable through standards harmonisation. In: 7th SaLTMiL Workshop on Creation and use of basic lexical resources for less-resourced languages, held in conjunction with LREC 2010, Valetta, Malta

Kemps-Snijders M, Windhouwer M, Wittenburg P, Wright S (2009) ISOcat: Remodelling metadata for language resources. International Journal of Metadata, Semantics and Ontologies 4(4):261–276

Kingsbury P, Palmer M (2002) From treebank to propbank. In: Proceedings of the 3rd International Conference on Language Resources and Evaluation (LREC-2002), Citeseer, pp 1989–1993

König E, Lezius W (2000) A description language for syntactically annotated corpora. In: Proc. 18th International Conference on Computational Linguistics (COLING 2000), Saarbrücken, Germany, pp 1056–1060

Leech G, Wilson A (1996) EAGLES recommendations for the morphosyntactic annotation of corpora. http://www.ilc.cnr.it/EAGLES/annotate/ annotate.html, version of March 1996

Lezius W (2002) TIGERSearch. Ein Suchwerkzeug für Baumbanken. In: Proceedings of the 6. Konferenz zur Verarbeitung natürlicher Sprache (6th Conference on Natural Language Processing, KONVENS 2002), Saarbrücken, Germany

Marcus M, Santorini B, Marcinkiewicz MA (1993) Building a large annotated corpus of English: the Penn Treebank. Computational Linguistics 19(2):313–330

Marcus M, Santorini B, Marcinkiewicz M (1994) Building a large annotated corpus of English: The Penn Treebank. Computational Linguistics 19(2):313–330

McCrae J, Spohr D, Cimiano P (2011) Linking lexical resources and ontologies on the semantic web with Lemon. The Semantic Web: Research and Applications pp 245–259

McCrae J, Montiel-Ponsoda E, Cimiano P (this vol.) Integrating WordNet and Wiktionary with *lemon*. pp 25–34

Müller C, Strube M (2006) Multi-level annotation of linguistic data with MMAX2. In: Corpus Technology and Language Pedagogy, Peter Lang, Frankfurt am Main, pp 197–214

Nordhoff S (this vol.) Linked Data for linguistic diversity research: Glottolog/Langdoc and ASJP. pp 191–200

Nuzzolese A, Gangemi A, Presutti V (2011) Gathering lexical linked data and knowledge patterns from framenet. In: Proceedings of the sixth international conference on Knowledge capture, ACM, pp 41–48

O'Donnell M (2000) Rsttool 2.4 – a markup tool for Rhetorical Structure Theory. In: Proc. International Natural Language Generation Conference (INLG'2000), Mitzpe Ramon, Israel, pp 253–256

Pareja-Lora A (this vol.) OntoLingAnnot's ontologies: Facilitating interoperable linguistic annotations (up to the pragmatic level). pp 117–127

Pareja-Lora A, Aguado de Cea G (2010) Ontology-based interoperation of linguistic tools for an improved lemma annotation in Spanish. In: Proceedings of LREC 2010, Valetta, Malta

Pustejovsky J, Meyers A, Palmer M, Poesio M (2005) Merging PropBank, Nom-Bank, TimeBank, Penn Discourse Treebank and Coreference. In: Proc. ACL Workshop on Frontiers in Corpus Annotation 2005, Ann Arbor, MI, USA

Rehm G, Eckart R, Chiarcos C (2007) An OWL-and XQuery-based mechanism for the retrieval of linguistic patterns from XML-corpora. In: Proc. RANLP 2007, Borovets, Bulgaria

Romary L, Zeldes A, Zipser F (2011) <tiger2/> Serialising the ISO SynAF syntactic object model. Arxiv preprint arXiv:1108.0631

Saulwick A, Windhouwer M, Dimitriadis A, Goedemans R (2005) Distributed tasking in ontology mediated integration of typological databases for linguistic research. In: Proc. 17th Conf. on Advanced Information Systems Engineering (CAiSE'05), Porto

Schiehlen M (2004) Optimizing algorithms for pronoun resolution. In: Proc. 20th International Conference on Computational Linguistics (COLING), Geneva, pp 515–521

Schmidt T (2004) EXMARaLDA – Ein System zur computergestützten Diskurstranskription. In: Mehler A, Lobin H (eds) Automatische Textanalyse. Systeme und Methoden zur Annotation und Analyse natürlichsprachlicher Texte, Verlag für Sozialwissenschaften, Wiesbaden, Germany, pp 203–218

Schmidt T, Chiarcos C, Lehmberg T, Rehm G, Witt A, Hinrichs E (2006) Avoiding data graveyards. In: Proceedings of the E-MELD workshop on Digital Language Documentation, East Lansing

Skut W, Krenn B, Brants T, Uszkoreit H (1997) An annotation scheme for free word order languages. In: Proc. 5th Conference on Applied Natural Language Processing (ANLP), Washington, D.C.

Skut W, Brants T, Krenn B, Uszkoreit H (1998) A linguistically interpreted corpus of German newspaper text. In: Proc. ESSLLI Workshop on Recent Advances in Corpus Annotation, Saarbrücken, Germany

Stede M, Bieler H, Dipper S, Suriyawongkul A (2006) Summar: Combining linguistics and statistics for text summarization. In: Proc. 17th European Conference on Artificial Intelligence (ECAI-06), Riva del Garda, Italy, pp 827–828

Trißl S, Leser U (2007) Fast and practical indexing and querying of very large graphs. In: Proc. 2007 ACM SIGMOD international conference on Management of data, pp 845–856. ACM

Windhouwer M, Wright SE (this vol.) Linking to linguistic data categories in ISO-cat. pp 99–107

Zeldes A, Ritz J, Lüdeling A, Chiarcos C (2009) ANNIS: A search tool for multi-layer annotated corpora. In: Proc. Corpus Linguistics, Liverpool, UK, pp 20–23

Zipser F, Romary L (2010) A model oriented approach to the mapping of annotation formats using standards. In: Proc. LREC-2010 Workshop on Language Resource and Language Technology Standards (LR&LTS 2010), Valetta, Malta

The German DBpedia: A Sense Repository for Linking Entities

Sebastian Hellmann, Claus Stadler, and Jens Lehmann

Abstract The modeling of lexico-semantic resources by means of ontologies is an established practice. Similarly, general-purpose knowledge bases are available, e.g. DBpedia, the nucleus for the Web of Data. In this section, we provide a brief introduction to DBpedia and describe recent internationalization efforts (including the creation of a German version) around it. With DBpedia serving as an entity repository it is possible to link the Web of Documents with the Web of Data via DBpedia identifiers. This function is provided by DBpedia Spotlight, which we briefly introduce. We then show how the NLP Interchange Format (NIF) can be used to represent this linking transparently for applications. Using the OLiA ontologies to represent linguistic annotations, NIF allows to represent the output of NLP tools, such as DBpedia Spotlight, in a uniform way.

1 DBpedia

DBpedia (Lehmann et al., 2009; Auer et al., 2008) is a community effort to extract structured information from Wikipedia and to make this information available on the Web. The main output of the DBpedia project is a data pool that (1) is widely used in academics as well as industrial environments, that (2) is curated by the community of Wikipedia and DBpedia editors, and that (3) has become a major crystallization point and a vital infrastructure for the Web of Data. DBpedia is one of the most prominent Linked Data examples and presently the largest hub in the Web of Linked Data (Fig. 1). The extracted RDF knowledge from the English Wikipedia is published and interlinked according to the Linked Data principles.

Sebastian Hellmann · Claus Stadler · Jens Lehmann
Universität Leipzig, Fakultät für Mathematik und Informatik, Abt. Betriebliche Informationssysteme, Johannisgasse 26, 04103 Leipzig, Germany e-mail: hellmann@informatik.uni-leipzig.de

C. Chiarcos et al. (eds.), *Linked Data in Linguistics*,
DOI 10.1007/978-3-642-28249-2_17, © Springer-Verlag Berlin Heidelberg 2012

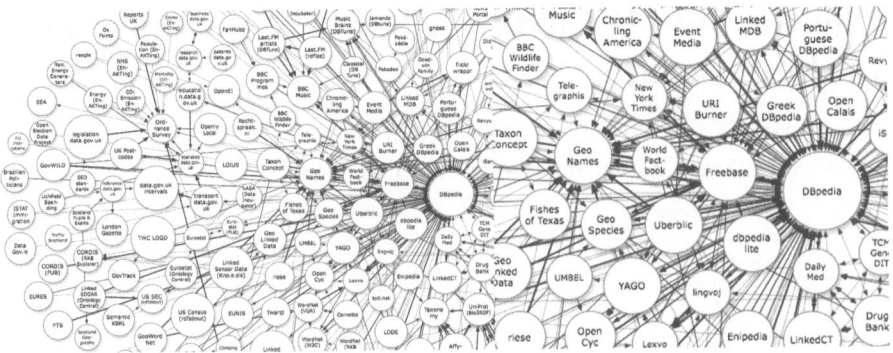

Fig. 1 Excerpt of the data sets interlinked with DBpedia. Source: http://lod-cloud.net (with kind permission of Anja Jentzsch and Richard Cyganiak)

According to Bizer (2011), the latest version of DBpedia (3.7 at the time of writing) contains more than 3.64 million things. Among them are 416,000 persons, 526,000 places, 106,000 music albums, 60,000 films, 17,500 video games, 169,000 organizations, 183,000 species and 5,400 diseases, which are classified in a consistent ontology. Multilingual labels and abstracts in about 97 different languages are provided. Furthermore, there are 2,724,000 links to images, 6,300,000 links to external web pages, 6,200,000 external links into other RDF datasets, and 740,000 Wikipedia categories. Approximately 385 million RDF triples were extracted from the English Wikipedia, whereas an additional 665 million ones originate from other language editions and links to external datasets. As such, the dataset amounts to roughly 1 billion triples.

Currently, the DBpedia Ontology is maintained in a crowd-sourcing approach and thus freely editable on a *Mappings Wiki*:[1] There contributors can 1) edit mapping rules about how Wikipedia infoboxes relate to OWL classes and properties, and 2) edit the ontology, such as adding class labels or modelling subsumption hierarchies. An example is shown in Fig. 2: For the article "Mount Everest", a corresponding DBpedia resource is created, which is classified as `dbp-owl:Mountain`. This is because the article contains the infobox "Infobox_Mountain", for which a contributor to the Mappings Wiki stated this rule. The key-value pairs used in the infobox are mapped to RDF triples via further mapping rules.

2 Internationalization of DBpedia

Early versions of the DBpedia Information Extraction Framework (DIEF) only used the English Wikipedia as their sole source. Yet, since then, its focus has shifted to build a fused dataset by integrating information from many different Wikipedia edi-

---

[1] http://mappings.dbpedia.org

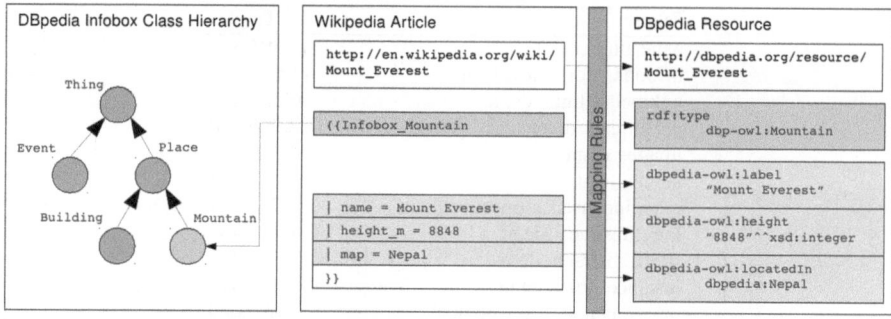

Fig. 2 Rule-based manipulation of extracted data in DBpedia Mappings Wiki

tions. The emphasis of this fused DBpedia is still on the English Wikipedia as it is the most abundant language edition. During the fusion process, however, language-specific information was lost or ignored. The aim of the current research in internationalization (Kontokostas et al., 2011) is to establish best practices (complemented by software) that allow the DBpedia community to easily generate, maintain and properly interlink language-specific DBpedia editions. In a first step, we realized a language-specific DBpedia version using the Greek Wikipedia as a basis for prototypical development (Kontokostas et al., 2011). Soon, however, the approach was generalized and applied to 15 other Wikipedia language editions (Bizer, 2011), amongst them the localized German DBpedia. The German Wikipedia is currently the second largest Wikipedia[2] with about 1.3 million articles. With the current version of DIEF, it is responsible for the third largest localized DBpedia with a total of 73 million RDF triples (German being the second largest one). The German community at the DBpedia Mappings Wiki has started to create mappings for infoboxes and achieved a coverage of about 52.89%.[3] A summary of the number of triples extracted for the German DBpedia, hosted by the Freie Universität Berlin,[4] is shown in Tab. 1.

Most importantly, DBpedia provides background knowledge for around 3.64 million entities (1.3 million in German) with a high stability with respect to the identifier-to-sense assignment (Hepp et al., 2007). This means that once a piece of text is correctly linked to DBpedia identifier representing the sense, it can be expected that this assignment remains stable.

---

[2] accessed on Dec 7th, 2011, `http://meta.wikimedia.org/wiki/List_of_Wikipe dias`

[3] accessed on Dec 8th, 2011, `http://mappings.dbpedia.org/server/statistics/ de/`

[4] `http://de.dbpedia.org`

Filename	Triples
page_links_de.nt	41.395.828
infobox_properties_de.nt	11.055.387
wikipedia_links_de.nt	6.539.427
persondata_de.nt	2.996.640
images_de.nt	2.329.617
labels_de.nt	2.180.233
mappingbased_properties_de.nt	1.764.684
long_abstracts_de.nt	1.137.481
short_abstracts_de.nt	1.137.481
instance_types_de.nt	804.913
interlanguage_links_de.nt	694.064
disambiguations_de.nt	636.481
pnd_de.nt	152.831
homepages_de.nt	117.856
specific_mappingbased_properties_de.nt	103.763
infobox_property_definitions_de.nt	21.510
geo_coordinates_de.nt	36
total	**73.068.232**

Table 1 Statistics of the extracted triples for the German DBpedia.

3 DBpedia Spotlight

One example application of DBpedia is DBpedia Spotlight (Mendes et al., 2011), a tool for annotating mentions of DBpedia resources in natural language, providing a solution for linking text to the Linked Open Data cloud through DBpedia. DBpedia Spotlight performs term extraction, maps terms to candidate resources and automatically selects a resource based on the context of the input text.

The most basic term extraction strategy in DBpedia Spotlight is based on a dictionary of known DBpedia resource names extracted from page titles, redirects and disambiguation pages. These names are shared in the DBpedia Lexicalization Dataset.[5] The graph of labels, redirects and disambiguations in DBpedia is used to extract a lexicon that associates multiple surface forms to a resource and interconnects multiple resources to an ambiguous name.

First, Wikipedia page titles can be seen as community-approved names. Further, redirects to URIs indicate synonyms or alternative surface forms, including common misspellings and acronyms. And finally, disambiguations provide ambiguous names that are "confusable" with all resources they link to. Their labels (after basic cleaning of words within parenthesis) become names for all target resources in the disambiguation page. In order to score the association between names and DBpedia resources, page links in Wikipedia are used. For each page link, one association between a name in the anchor text with the resource in the target page is counted. Based on this statistics, a number of scores have been derived and shared in the DBpedia Lexicalization dataset.

---

[5] http://wiki.dbpedia.org/Lexicalizations

Other term extraction techniques already available through DBpedia Spotlight include Keyphrase Extraction (Frank et al., 1999), a non-lexicalized segmentation approach using a shallow parser, and Named Entity Recognition of People, Locations and Organizations based on OpenNLP.

The disambiguation strategy used in DBpedia Spotlight 0.5 models each DBpedia resource in a vector space model of words extracted from Wikipedia paragraphs containing page links. A paragraph is added to the context of a DBpedia resource if that resource's corresponding Wikipedia page is the target of a page link in that paragraph. The words in the paragraph are further scored based on the TF*ICF measure (Mendes et al., 2011), which scores words on their ability to distinguish between known senses of a given term.

DBpedia Spotlight allows users to configure the annotation behaviour based on a number of parameters. Through the use of the DBpedia Ontology, the system allows users to restrict the annotations to a set of classes (e.g. Politician, Restaurants, etc.), or to any arbitrary query expressed in SPARQL. Furthermore, a number of scores, including prominence, contextual pertinence and confidence allows users to deal with the natural trade-off between precision and recall (Gordon and Kochen, 1989).

One recent development is the internationalization of DBpedia Spotlight and the development of entity disambiguation services for German and Korean has begun. Other languages will follow soon including the evaluation of the performance of the algorithms in other languages.

4 NLP Interchange Format

The NLP Interchange Format (NIF)[6] brings together the previously described concepts. It is an RDF/OWL-based format that aims to achieve interoperability between Natural Language Processing (NLP) tools, language resources and annotations. The core of NIF consists of a vocabulary, which can represent strings as RDF resources. A special URI design is used to pinpoint annotations to a part of a document. These URIs can then be used to attach arbitrary annotations to the respective character sequence. Based on these URIs, annotations can be interchanged between different NLP tools and applications. NIF consists of 3 components:

1. Structural Interoperability (cf. Sect. 4.1): URI recipes are used to anchor annotations in documents with the help of fragment identifiers. The URI recipes are complemented by two ontologies (String Ontology and Structured Sentence Ontology), which are used to describe the basic types of these URIs (String, Document, Word, Sentence) as well as the relations between them (sub/super string, next/previous word).

2. Conceptual Interoperability: Best practices for annotating these URIs are given to provide interoperability. OLiA,[7] as presented in Chiarcos (this vol.), is used

---

[6] Specification 1.0: http://nlp2rdf.org/nif-1.0

[7] http://purl.org/olia

for the grammatical features, while the SCMS Vocabulary[8] and DBpedia are used for sense tagging.

3. Access Interoperability: An interface description for NIF Components and Web Services allows NLP tools to interact on a programmatic level.

4.1 Anchoring Web Annotations

One basic use case of NIF is to allow NLP tools to exchange annotations about (Web) documents in RDF. The first prerequisite for achieving this is that strings in documents can be referred to by URIs, so they can be used as a subject in RDF triples. A quite simple example is depicted in Fig. 3, which can be addressed with two possible URI recipes according to the NIF 1.0 specification:

1. A URI scheme with offsets can be used, which is in general easy to compute and handle programmatically:

```
http://www.depardieu.8m.com/#offset_22295_22304_Depardieu
```

Note that the HTML document is treated as a string or a character sequence. The # is used in this case to address a fragment of the whole document, hence the naming "fragment identifier". 22295 denotes the offset of the first character in the (source code of) the website, whereas 22304 is the last character. "Depardieu" is an addition to make the URI more readable.

2. A URI scheme based on the context and md5 hashes, which is more stable compared to the previous recipe, could be used:

```
http://www.depardieu.8m.com/#hash_6_9_
e7146a74239c3878aedf0c45c6276618\_Depardieu
```

Considering the example "nbsp; (Depardieu) is on", here the context is 6 characters before and after the occurrence of the substring in question. A hash from this context can be computed and added to the URI. The larger the context, the more likely it is that two different strings are not assigned the same URI. The advantage of hashes is that URIs may stay valid even when the document changes.

A NIF 1.0 model, which links the second occurrence of *Depardieu* to the French DBpedia could contain the following RDF triples:

```
@prefix :        <http://www.depardieu.8m.com/#>
@prefix fr:      <http://fr.dbpedia.org/resource>
@prefix str:     <http://nlp2rdf.lod2.eu/schema/string/> .
@prefix scms:    <http://ns.aksw.org/scms/> .

:offset_22295_22304_Depardieu
    scms:means  fr:Gerard_Depardieu ;
    rdf:type    str:OffsetBasedString .
```

---

[8] http://scms.eu

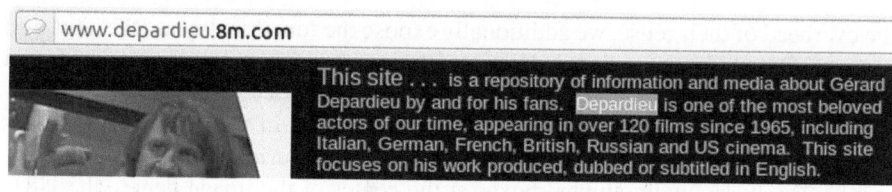

Fig. 3 The second occurence of `Depardieu` is highlighted and is linked in the example with the French DBpedia resource about Gérard Depardieu. Source: http://www.depardieu.8m.com/

```
:offset_0_54093_%3C!doctype%20html%20publi
    rdf:type      str:OffsetBasedString ;
    rdf:type      str:Document ;
    str:subString :offset_22295_22304_Depardieu;
    str:sourceUrl <http://www.depardieu.8m.com/> .
```

Similarly, the output of NLP tools can be represented, e.g., by associating *Depardieu* with its language (e.g., a Glottolog or lexvo identifier), with a syntactic parse tree (e.g., as specified in POWLA), or with morphosyntactic annotations (as provided by OLiA).

4.2 Integration of NLP Tools Using NIF

Many existing NLP tools can be integrated with NIF by providing a wrapper that serializes and deserializes between NIF and its native format. Our experience is, that the implementation efforts of such wrappers is limited, thereby resulting in low integration costs. An NLP pipeline can then be formed by either exchanging NIF models directly, or by merging the NIF models generated by tools that take documents as input.

The NIF wrappers and the special URI recipes ensure that equal strings in equal contexts are assigned equal URIs. Therefore, the integration of the outputs of NLP tools providing annotations for the same part of a document can be done by simply merging the NIF models.

NIF Webservices

We have implemented NIF wrappers for Stanford Core NLP,[9] Apache OpenNLP.[10] Snowball Stemmer,[11] DBpedia Spotlight, and Monty Lingua.[12] In order to minimize

---

[9] http://nlp.stanford.edu/software/corenlp.shtml
[10] http://incubator.apache.org/opennlp
[11] http://snowball.tartarus.org/
[12] http://web.media.mit.edu/~hugo/montylingua/

the overhead of their reuse, we additionally expose the functionality as RESTful web services. Detailed information and documentation about these services can be found at the NLP2RDF website.[13] Online demos are also provided.[14]

Figure 4 shows an example about the merged output of the Snowball Stemmer together with Stanford Core for the sentence "My favourite actor is Natalie Portman!". The horizontally aligned boxes at the center of the image denote the URIs referencing the individual substrings of the source sentence, whereas the attached annotations originate from the Snowball stemmer (arrows labelled `Stem`), and the Stanford framework (`Pos Tag`, `Type`, `Lemma`). Additionally, the model is enriched with NIF metadata, such as the successor relation between the constituent words of the sentence (not shown in this figure).

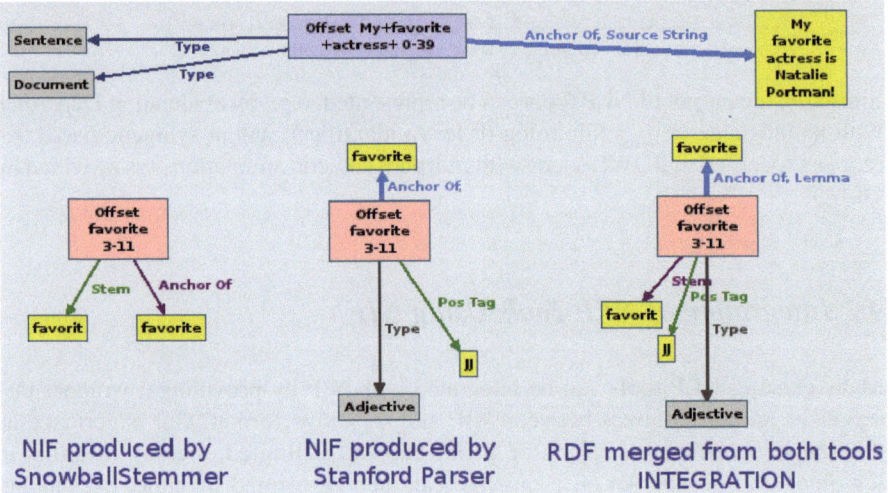

Fig. 4 Merged output of Snowball Stemmer and Stanford.

5 Summary

We have presented several technologies, which together show how text can be annotated using background knowledge from the Linked Open Data cloud:

- DBpedia as cross domain knowledge base.
- Different language editions of DBpedia, in particular the German DBpedia.
- DBpedia Spotlight, as a tool for annotating mentions of DBpedia entities in text.
- NIF as format for interchanging annotations.

---

[13] http://nlp2rdf.org

[14] http://nlp2rdf.lod2.eu/demo.php

Overall, we argue that the (German) DBpedia can become a sense repository for annotating entities. In future work, we will also look at the integration of special entities through knowledge bases like LinkedGeoData (Auer et al., 2009; Stadler et al., 2011). NIF provides an easy way to reuse resources from the LOD cloud in general and allows a seamless integration of several NLP tools.

Acknowledgements We would like the thank Pablo Mendes for his contributions to this work.

References

Auer S, Bizer C, Kobilarov G, Lehmann J, Cyganiak R, Ives Z (2008) DBpedia: A nucleus for a web of open data. In: Proceedings of the 6th International Semantic Web Conference (ISWC). Springer, Lecture Notes in Computer Science, vol 4825, pp 722–735, DOI 10.1007/978-3-540-76298-0_52

Auer S, Lehmann J, Hellmann S (2009) LinkedGeoData - adding a spatial dimension to the web of data. In: Proc. of 8th International Semantic Web Conference (ISWC), DOI 10.1007/978-3-642-04930-9_46, URL http://www.informatik.uni-leipzig.de/~auer/publication/linkedgeodata.pdf

Bizer C (2011) Dbpedia 3.7 released, including 15 localized editions. http://blog.dbpedia.org/2011/09/11/dbpedia-37-released-including-15-localized-editions/

Chiarcos C (this vol.) Interoperability of corpora and annotations. pp 161–179

Frank E, Paynter GW, Witten IH, Gutwin C, Nevill-Manning CG (1999) Domain-specific keyphrase extraction. In: Proceedings of the Sixteenth International Joint Conference on Artificial Intelligence. Morgan Kaufmann Publishers Inc., San Francisco, CA, USA, IJCAI '99. pp 668–673, URL http://dl.acm.org/citation.cfm?id=646307.687591

Gordon M, Kochen M (1989) Recall-precision trade-off: A derivation. Journal of the American Society for Information Science 40(3):145–151. URL http://www3.interscience.wiley.com/cgi-bin/jtoc/27981/

Hepp M, Siorpaes K, Bachlechner D (2007) Harvesting wiki consensus: Using wikipedia entries as vocabulary for knowledge management. IEEE Internet Computing 11(5):54–65. URL http://dblp.uni-trier.de/rec/bibtex/journals/internet/HeppSB07

Kontokostas D, Bratsas C, Auer S, Hellmann S, Antoniou I, Metakides G (2011) Towards linked data internationalization - realizing the Greek DBpedia. In: Proceedings of the ACM WebSci'11

Lehmann J, Bizer C, Kobilarov G, Auer S, Becker C, Cyganiak R, Hellmann S (2009) DBpedia - a crystallization point for the web of data. Journal of Web Semantics 7(3):154–165, DOI 10.1016/j.websem.2009.07.002, URL http://jens-lehmann.org/files/2009/dbpedia_jws.pdf

Mendes PN, Jakob M, García-Silva A, Bizer C (2011) DBpedia Spotlight: Shedding light on the web of documents. In: Proc. 7th International Conference on Semantic Systems (I-Semantics)

Stadler C, Lehmann J, Höffner K, Auer S (2011) Linkedgeodata: A core for a web of spatial open data. Semantic Web Journal

Linked Data for Linguistic Diversity Research: Glottolog/Langdoc and ASJP Online

Sebastian Nordhoff

Abstract Most of the linguistic resources available to day are about the world's major languages. This paper discusses two projects which have world-wide coverage as their aim. Glottolog/Langdoc is an attempt to attain near-complete bibliographical coverage for the world's lesser-known languages (i.e. 95% of the world's linguistic diversity), which then provides solid empirical ground for extensional definitions of languages and language classification. Automated Similarity Judgment Program (ASJP) online provides standardized lexical distance data for 5800 languages from Brown et al. (2008) as Linked Data. These two projects are the first attempt at a Typological Linked Data Cloud, to which PHOIBLE by Moran (this vol.) can easily be added in the future.

1 Introduction

The original motivation underlying the development of standards such as RDF has been to describe resources, e.g. books in a library. The Glottolog/Langdoc project exemplifies a similar application scenario for the linguistic domain, i.e. the collection and formalization of **information about languages and language resources** within the Linked Open Data cloud. By doing so, Glottolog/Langdoc covers the band-width of languages in the world as far as possible, i.e. with a certain emphasis – albeit not a strict focus – on less-resourced languages.

Section 2 gives an overview of the bibliographical part of the project (Langdoc), Sect. 3 introduces the notion of **languoid**, a data structure for the modeling of genealogical relationships between language families, languages and dialects (Glottolog), Sect. 4 summarizes the resource types provided for the Linguistic Linked

Sebastian Nordhoff
Department of Linguistics, Max Planck Institute for Evolutionary Anthropology, Deutscher Platz 6, 04103 Leipzig, Germany e-mail: sebastian_nordhoff@eva.mpg.de

C. Chiarcos et al. (eds.), *Linked Data in Linguistics*,
DOI 10.1007/978-3-642-28249-2_18, © Springer-Verlag Berlin Heidelberg 2012

Open Data cloud, and Sect. 5 illustrates a concrete application of Glottolog/Langdoc in the context of the related ASJP project.

2 Multilingual References on a World-wide Scale: Langdoc

Linguistic resources can be classified into resources *for* a language (dictionaries, thesauri, spellcheckers etc) and resources *about* a language (descriptive grammars, description of the history of a language, theoretical linguistic analysis of the phonology/morphology/syntax of a language). Linked Open Data has made good progress in the former area (WordNet etc), but has not really started yet in the latter. The Langcoc project aims at remedying this by providing near-complete bibliographical information about the world's lesser known languages as Linked Data.

2.1 Lesser-Known Languages

Current estimates suggest that there are about 7000 different languages spoken on Earth. As far as the amount of resources available for a language is concerned, there is a clear split. One the one hand, we have high resource languages. Those are national languages (e.g. Swedish), or other languages with a long written tradition (Catalan). These languages provide commercial viability for linguistic resources. On the other hand, we have languages with a very short written tradition, or even no written tradition at all. For those languages, there is no commercial interest in providing linguistic resources. The resources treating those languages are academic or missionary. The Langdoc project focuses on the latter group. I will call this group of languages 'lesser known languages'. Other names for the same group are 'low-density languages', and 'low resource languages'. Our current estimate is that there are only about 200 'better known languages' and the remainder, i.e. 6800 languages, have to be considered lacking in terms of resources.

2.2 Resource Collection and Resource Collections

A number of dedicated individuals have consecrated a lot of their time to collecting references about lesser-known languages. Alain Fabre (Fabre, 2005) has collected 26 634 references, treating 615 languages of South America; Jouni Maho (Maho, 2001) has collected 59 788 references treating 1 994 languages of Africa. *SIL international* have collected 6 246 references for Papua New Guinea (410 languages), in addition to the 18 190 references they provide on a world-wide scale. The Alaska

Native Center[1] provides 13 876 references for the languages of Alaska. The coverage of these bibliographies can be considered near-complete. For the other areas of the world, there are no comparable bibliographies. We have to rely on the aggregation of a number of large world-wide bibliographies and a number of smaller areal bibliographies[2] and hope that they will complete each other. It is likely that the more obscure references are currently not included in our resource collection for those areas of the world.

Langdoc lists 166 459 resources providing information about the world's linguistic diversity.[3] The resources are tagged for resource type (grammar, word list, text collection etc), macroarea (geographic region), and language. Table 1 gives an overview of the resources covered so far, classified by macroarea and document type.

Table 1 Language resources in Langdoc according to geographic region and document type

area	refs	document type	refs	document type	refs
Africa	74 787	comparative treatise	13 827	phonology	1 942
South America	32 897	grammar sketch	13 810	bibliography	1 464
Eurasia	16 879	ethnographic treatise	13 504	specific feature	1 362
Pacific	15 424	grammar	10 209	text	1 039
Australia	7 557	overview	8 273	sociolinguistics	943
North America	3 815	dictionary	7 408	dialectology	797
Middle America	1 897	wordlist	5 552	new testament	143

Our plan is to significantly expand the coverage of Langdoc in the years to come. The ingestion of future resources is guided by the following two principles, where the first has a higher priority than the second:

---

[1] http://www.uaf.edu/anla/

[2] ASJP Automated Similarity Judgment Program bibliography http://lingweb.eva.mpg.de/asjp/index.php/ASJP; Alain Fabre's "Diccionario etnolingüístico y guía bibliográfica de los pueblos indígenas sudamericanos" http://www.tut.fi/fabre/BookIntervetVersio; The bibliography of the Papua New Guinea branch of SIL http://www.sil.org/pacific/png/; Randy LaPolla's Tibeto-Burman bibliography http://victoria.linguistlist.org/~lapolla/bib/index.htm; The bibliography of the South Asian Linguistics Archive http://www.sealang.net/library/; Frank Seifart's bibliography www.eva.mpg.de/lingua/staff/seifart.html; The World Atlas of Language Structures www.wals.info; Harald Hammarström's bibliography http://haraldhammarstrom.ruhosting.nl/; The catalogue of the Max Planck Institute for Evolutionary Anthropology in Leipzig, www.eva.mpg.de/library; The SIL bibliography www.ethnologue.com/bibliography.asp; The web-version of EBALL, by Jouni Maho and Guillaume Ségerer http://sumale.vjf.cnrs.fr/Biblio/; Jouni Maho's bibliography of Africa; http://goto.glocalnet.net/maho/eball.html; Tom Güldemann's bibliography of Africa http://www2.hu-berlin.de/asaf/Afrika/Mitarbeiter/Gueldemann.html; Chintang-Puma Documentation Project http://www.uni-leipzig.de/~ff/cpdp/

[3] Note that the we only provide the reference, but no copy of the work itself. We link to WorldCat, GoogleBooks and Open Library to help users retrieve a copy.

1. For every language, provide a reference of the most extensive piece of documentation.
2. Beyond that, provide as many references as possible

2.3 Storage and Retrieval

All references are stored in a relational database. A web frontend provides access to the data. The references are retrievable via standard bibliographical fields such as author, year, title, etc. Additionally, Langdoc allows for for genealogical searches in a step-free manner. This is accomplished by using a set-theoretic approach: English is a subset of Germanic, and a subset of Indo-European. This means that a reference associated with English is associated with Germanic (and Indo-European) at the same time.

A researcher interested in languages of the Pacific Ocean could search at any level of the deeply nested tree of Austronesian languages (Fig. 1). Queries like 'Give me any dictionary of an Oceanic language' or 'Give me any grammar of a Polynesian language' become possible. The genealogical data just mentions lead us to the counterpart of Langdoc: Glottolog.

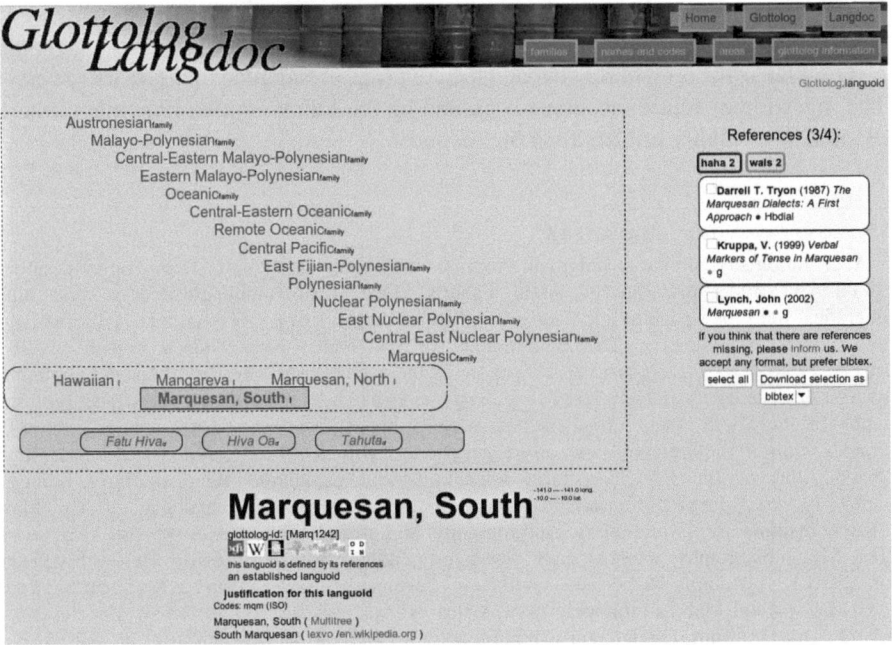

Fig. 1 The page for the languoid 'Marquesan' with the genealogy on the left and references on the right.

3 A Resource-Based Definition of Languages: Glottolog

The traditional approach to language classification has an intensional approach: languages have an essence. The problem is that this essence is not accessible to computers, so that models relying on the essence of a particular language are difficult to implement. To circumvent this issue, we use a novel approach in that we use an *extensional* definition of languages: a language is defined by the set of documents which describe it (Nordhoff and Hammarström, 2011). This has two advantages: we can exploit the available understanding of the ways how to model documents and associated metadata, and languages without documentation disappear from the model space. While it would of course be desirable to have information about any and all languages ever spoken by humans, it is a fact that we do not have all this information, and the scientific method dictates that we stick to what is observable. In this sense, the disappearance of languages without empirical attestation from our model is actually an advantage. This does not mean that those languages would be less interesting; it simply means that they are not yet part of the observed universe of Western academia.

Another advantage is that relations between languoids can be modeled in a set theoretic fashion. Let $\mathfrak{S}$ be the set of all documents treating Swedish, $\mathfrak{D}$ be the set of all documents treating Danish and $\mathfrak{N}$ be the set of all documents treating Norwegian. The union of these sets is then the set of all documents treating a North Germanic language. This subset relation can be iterated up to the root node 'Indo-European'.

This set-theoretic approach allows for a step-free modeling of language classifications. It furthermore does away with the need of singling out a particular level 'language' (as compared to dialect or variety), which short-circuits eternal discussions as to whether variety X is a language or a dialect. All sets have persistent IDs, which can be used to uniquely refer to a set, and no IDs are privileged.

The provision of URIs for documents and sets means that conflicting opinions can be modeled: Some researchers for instance assume that Baltic and Slavic are direct children of Indo-European while others assume an intervening node Balto-Slavic. The unique URIs allow researchers disagreeing about this aspect of the classification to state that the lower parts of the tree would still be identical. The document-centric approach furthermore allows the inference that Baltic, Slavic, and Indo-European still have the same meaning in both classifications, based on the extensional definition based on documents. The only differences is that in the second classification, there will be an additional languoid (more on this term below) 'Balto-Slavic' with associated documents, which is entirely missing from the first classification.

This leads to the modeling employed by Glottolog. As stated above, we employ a set-theoretic approach. Every languoid is seen as a set. Subset and superset relations can model genealogical relationships. In this particular case, Glottolog employs skos:narrower and skos:broader to model the relation between a larger languoid like North Germanic and a smaller languoid like Danish. Treating languages as concepts allows to make use of general insights gained in other areas where taxonomies of concepts are used. At the same time, this means that Glottolog/Langdoc languoids are not comparable to languages in the sense of GOLD (Farrar and Langendoen, 2003a) or Ethnologue (Lewis, 2009b). The former are concepts (as defined in SKOS) while the latter are linguistic systems (as defined in Dublin Core).

4 Glottolog/Langdoc and Linked Data

Glottolog/Langdoc provides two types of resources as Linked Open Data: languoids and bibliographical records.

Languoid is a cover term for dialect, language, and language family (Good and Hendryx-Parker, 2006). Every languoid has its own URI and is annotated for ancestors, siblings, children, names, codes, geographic location and references. Links are provided to Multitree,[4] LL-Map (Xie et al., 2009), LinguistList,[5] Ethnologue (Lewis, 2009a), ODIN (Lewis, 2006), WALS (Dryer, 2005), OLAC (Bird and Simons, 2001), Lexvo (de Melo and Weikum, 2008), and Wikipedia. Languoids are modeled using SKOS and RDFS and linked to ontologies like GOLD (Farrar and Langendoen, 2003b), Lexvo, and geo.[6]

Bibliographical records of **Language resources** are available in XHTML and RDF. Resources make use of Dublin Core (Weibel et al., 1998) and are annotated for the languoids they are applied to. Additionally, resources are linked to WorldCat,[7] GoogleBooks,[8] and Open Library.[9]

The database currently covers 166 459 resources and 94 049 languoids. Tables 1 and 2 provide an overview of its content according to different criteria.

The data collected within the Glottolog/Langdoc project are different from other linguistic data and cannot be linked according to the standard principles (e.g. *lemon*, see McCrae this volume). There are three nodes in the Linked Data cloud Glottolog/Langdoc can be hooked onto: ISBN numbers can link Glottolog/Langdoc to various resources, ISO 639-3 codes can link Glottolog/Langdoc to lexvo, and the provided geocodes can link to geographical data repositories like WGS84.[10]

---

[4] http://multitree.linguistlist.org

[5] http://linguistlist.org

[6] http://www.w3.org/2003/01/geo/wgs84_pos

[7] http://www.worldcat.org

[8] http://books.google.com

[9] http://openlibrary.org

[10] http://www.w3.org/2003/01/geo/wgs84_pos

language	refs	language	refs	language	refs
Swahili	1 916	Igbo	550	16 languages	300-399
Hausa	1 609	Sotho, Southern	539	31 languages	200-299
Nama	1 288	Arabic, Algerian	526	159 languages	100-199
Zulu	1 060	Oromo, Borana-Arsi-Guji	516	389 languages	50-99
Arabic, South Levantine	1 033	Turkish	511	647 languages	25-49
Yoruba	925	Tarifit	505	611 languages	15-24
Kabyle	897	Nyanja	504	533 languages	10-14
Thai	745	Arabic, Tunisian	498	1033 languages	5-9
Pulaar	743	Tachelhit	490	351 languages	4
Xhosa	739	Wolof	487	436 languages	3
Akan	729	Tibetan	483	612 languages	2
Éwé	713	Sotho, Northern	467	1045 languages	1
Tswana	703	Aymara, Central	462		
Mapudungun	610	Aymara, Southern	454		
Shona	597	Vietnamese	439		
Somali	591	Paraguayan Guaraní	436		
Amharic	554	Singa	405		

Table 2 Documentation status of the languages in Langdoc (excluding resource-heavy languages like English or German). The relative overrepresentation of African and South American languages is due to the extent of bibliographical coverage found in the work of Maho (2001) and Fabre (2005).

5 Lexical Distances as Linked Data Resources: ASJP Online

Having described the Glottolog/Langdoc resources, we will now turn to an example application that integrates these resources with lexical-semantic resources provided by another project, ASJP (Holman et al., 2011).

Genetic relatedness between languages can be established by comparing basic vocabulary from the languages under discussion in order to see whether cognate sets and corresponding sound changes can be found. This has been a very laborious task for a very long time. The Automated Similarity Judgment Program (ASJP) automatizes this task by providing standardized word lists for 5 395 languages according to a particular abstract phonetic representation (Brown et al., 2008). Using a distance metric like Levenshtein, the relative lexical distance between two languages can be computed. The results can be compiled in a distance matrix and be represented as a tree. Figure 2 shows a screenshot illustrating this for Slavic languages.

The ASJP website currently also allows on-the-fly clustering and dendrogram generation of custom sets of languages. A researcher interested in the classification of Basque could for instance compute a tree of a candidate family and Basque to see how Basque fits into this family as far as its basic vocabulary is concerned.

The resources of this project are currently being made available as Linked Data in RDF[11] including language names, ISO 639-3 codes, WALS codes, number of speakers, date of extinction (if applicable), longitude, latitude, 40-items word list, and lexical distance between any two languages. Language names, codes, and geo-

---

[11] http://cldbs.eva.mpg.de/asjp

ASJP
languages in family SLAVIC

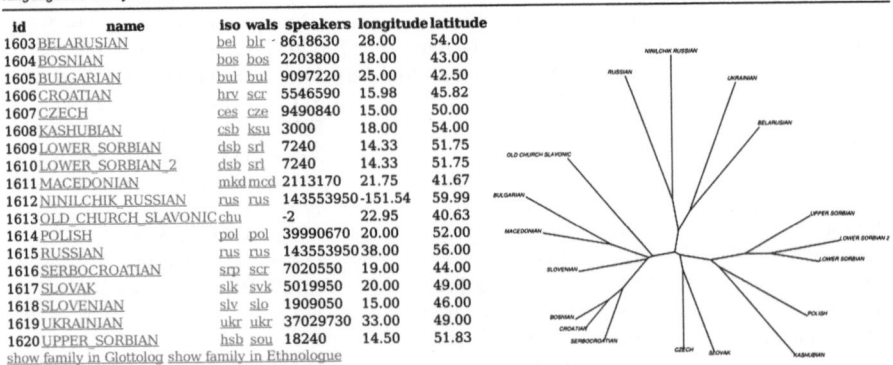

id	name	iso	wals	speakers	longitude	latitude
1603	BELARUSIAN	bel	blr	8618630	28.00	54.00
1604	BOSNIAN	bos	bos	2203800	18.00	43.00
1605	BULGARIAN	bul	bul	9097220	25.00	42.50
1606	CROATIAN	hrv	scr	5546590	15.98	45.82
1607	CZECH	ces	cze	9490840	15.00	50.00
1608	KASHUBIAN	csb	ksu	3000	18.00	54.00
1609	LOWER_SORBIAN	dsb	srl	7240	14.33	51.75
1610	LOWER_SORBIAN_2	dsb	srl	7240	14.33	51.75
1611	MACEDONIAN	mkd	mcd	2113170	21.75	41.67
1612	NINILCHIK_RUSSIAN	rus	rus	143553950	-151.54	59.99
1613	OLD_CHURCH_SLAVONIC	chu		-2	22.95	40.63
1614	POLISH	pol	pol	39990670	20.00	52.00
1615	RUSSIAN	rus	rus	143553950	38.00	56.00
1616	SERBOCROATIAN	srp	scr	7020550	19.00	44.00
1617	SLOVAK	slk	svk	5019950	20.00	49.00
1618	SLOVENIAN	slv	slo	1909050	15.00	46.00
1619	UKRAINIAN	ukr	ukr	37029730	33.00	49.00
1620	UPPER_SORBIAN	hsb	sou	18240	14.50	51.83

show family in Glottolog   show family in Ethnologue

Fig. 2 The ASJP page for Slavic languages. A list of the languages on the left is complemented by an unrooted tree showing the lexical proximity of the Slavic languages. We can distinguish East Slavic languages on top, South Slavic Languages to the left, and West Slavic languages to the right and at the bottom.

graphical coordinates allow for the integration of ASJP data into the Linked Open Data cloud. For instance, Glottolog makes use of an ASJP as a provider for dendrograms for arbitrary nodes.

ASJP links to the following other language projects (Glottolog/Langodoc, Ethnologue, OLAC, Multitree, LL-MAP, lexvo) via dcmi:relation. Further research will be necessary to provide a predicate more information than dcmi:relation. ASJP has incoming links from Glottolog/Langdoc.

ASJP provides lexical information in a wider sense. This information can in principle be linked to other lexical semantic resources like WordNet. This is however complicated by the particular phonetic representation ASJP uses as a storage format. This representation makes use of a limited set of characters. For instance, the German word *Knochen* 'bone' is represented as <knoX3n>. More serious is the reduction of phonological oppositions. As an example, roundedness is never recorded in ASJP lexemes. This means that French *deux* 'two' is represented in ASJP as <de>, the same representation *dé* 'dice' would receive. The conversion from Standard French phonology to ASJP representation is straightforward: /ø/ and /e/ both translate to <e>. The inverse direction is more problematic: <e> corresponds to both /ø/ and /e/, and it is not clear how an ASJP string like <de> should be translated. This phonological underspecification means that automated linking to other language-particular lexical semantic resources is severely hampered.

The third and last type of data are the lexical distances computed for pairs of languages. These distances are made available as floating point numbers. This means that third party projects can use these data for other purposes, e.g. refinement of the clustering algorithm or similar.

References

Bird S, Simons G (2001) The olac metadata set and controlled vocabularies. In: Proceedings of the ACL 2001 Workshop on Sharing Tools and Resources - Volume 15, Association for Computational Linguistics, Stroudsburg, PA, USA, STAR '01, pp 7–18. DOI 10.3115/1118062.1118065. Online version http://www.language-archives.org.

Brown CH, Holman EW, Wichmann S, Velupillai V (2008) Automated classification of the world's languages: A description of the method and preliminary results. STUF 61(4):286–308

Dryer MS (2005) Genealogical language list. In: Comrie B, Dryer MS, Gil D, Haspelmath M (eds) World Atlas of Language Structures, Oxford University Press, pp 584–644

Fabre A (2005) Diccionario etnolingüístico y guía bibliográfica de los pueblos indigenas sudamericanos. Book in Progress at http://butler.cc.tut.fi/~fabre/BookInternetVersio/Alkusivu.html accessed May 2005.

Farrar S, Langendoen D (2003a) A linguistic ontology for the semantic web. Glot International 7(3):97–100

Farrar S, Langendoen D (2003b) Markup and the GOLD ontology. In: EMELD Workshop on Digitizing and Annotating Text and Field Recordings, Michigan State University

Good J, Hendryx-Parker C (2006) Modeling contested categorization in linguistic databases. In: Proceedings of the EMELD Workshop on Digital Language Documentation, East Lansing, Michigan

Holman EW, Brown CH, Wichmann S, Müller A, Velupillai V, Hammarström H, Sauppe S, Jung H, Bakker D, Brown P, Belyaev O, Urban M, Mailhammer R, List JM, Egorov D (2011) Automated dating of the world's language families based on lexical similarity. Current Anthropology 52:841–875

Lewis M (ed) (2009a) Ethnologue: Languages of the World, Sixteenth edition. SIL International, Dallas, online version available at http://www.ethnologue.com/. Accessed on 2011-11-27.

Lewis MP (ed) (2009b) Ethnologue: Languages of the World, 16th edn. SIL, Dallas.

Lewis WD (2006) Odin: A model for adapting and enriching legacy infrastructure. In: Proceedings of the e-Humanities Workshop, held in cooperation with e-Science 2006: 2nd IEEE International Conference on e-Science and Grid Computing, Amsterdam, URL http://faculty.washington.edu/wlewis2/papers/ODIN-eH06.pdf, online version available at http://www.csufresno.edu/odin/

Maho J (2001) African Languages Country by Country: A Reference Guide, Göteborg Africana Informal Series, vol 1, 5th edn. Department of Oriental and African Languages, Göteborg University

de Melo G, Weikum G (2008) Language as a foundation of the Semantic Web. In: Bizer C, Joshi A (eds) Proceedings of the Poster and Demonstration Session at the 7th International Semantic Web Conference (ISWC 2008), CEUR, Karlsruhe, Germany, CEUR WS, vol 401

Moran S (this vol.) Using Linked Data to create a typological knowledge base. pp
    129–138

Nordhoff S, Hammarström H (2011) Glottolog/Langdoc: Defining dialects, lan-
    guages, and language families collections of resources. In: Proceedings of ISWC
    2011, URL http://iswc2011.semanticweb.org/fileadmin/iswc
    /Papers/Workshops/LISC/nordhoff.pdf

Weibel S, Kunze J, Lagoze C, Wolf M (1998) RFC 2413 - Dublin Core metadata for
    resource discovery. http://www.isi.edu/in-notes/rfc2413.txt

Xie Y, Aristar-Dry H, Aristar A, Lockwood H, Thompson J, Parker D, Cool B
    (2009) Language and location: Map annotation project - a gis-based infrastruc-
    ture for linguistics information management. In: Computer Science and Infor-
    mation Technology, 2009. IMCSIT '09. International Multiconference on, pp
    305–311, DOI 10.1109/IMCSIT.2009.5352710, URL http://ieeexplore.
    ieee.org/stamp/stamp.jsp?tp=&arnumber=5352710, online ver-
    sion at http://www.llmap.org

Linking Linguistic Resources: Examples from the Open Linguistics Working Group

Christian Chiarcos, Sebastian Hellmann, and Sebastian Nordhoff

Abstract The contributions of this part have described recent activities of the OWLG as a whole and of individual OWLG members aiming to provide linguistic resources as Linked Data. Here, we describe how linguistic resources can be linked with each other, and we illustrate possible use cases of information integration from various sources with example queries for the major types of linguistic resources: Using DBpedia (Hellmann et al., this vol.) to represent lexical-semantic resource, the German NEGRA corpus in its POWLA representation (Chiarcos, this vol.) to represent linguistic corpora, the OLiA ontologies (Chiarcos, this vol., Sect. 3) to represent repositories of linguistic terminology, and languoid definitions in Glottolog/Langdoc (Nordhoff, this vol.) to represent linguistic knowledge bases and metadata repositories.

We use data from German for illustration purposes, the NEGRA corpus, a linguistically annotated collection of newspaper articles from the Frankfurter Rundschau. The architecture described here, is, however, not specific to German, but can also be applied to other languages. In fact, the Glottolog/Langdoc resources have been developed primarily for language documentation and typological studies, but their applications as described below naturally extends for less-resourced languages.

Christian Chiarcos
Information Sciences Institute, University of Southern California, 4676 Admiralty Way # 1001, Marina del Rey, CA 90292 e-mail: chiarcos@daad-alumni.de

Sebastian Hellmann
Universität Leipzig, Fakultät für Mathematik und Informatik, Abt. Betriebliche Informationssysteme, Johannisgasse 26, 04103 Leipzig, Germany e-mail: hellmann@informatik.uni-leipzig.de

Sebastian Nordhoff
Department of Linguistics, Max Planck Institute for Evolutionary Anthropology, Deutscher Platz 6, 04103 Leipzig, Germany e-mail: sebastian_nordhoff@eva.mpg.de

C. Chiarcos et al. (eds.), *Linked Data in Linguistics*,
DOI 10.1007/978-3-642-28249-2_19, © Springer-Verlag Berlin Heidelberg 2012

1 The NEGRA Corpus

This contribution illustrates possible use cases of information integration from various sources, taking data from the NEGRA Corpus as an example.[1] The NEGRA corpus consists of 355,096 tokens (20,602 sentences) of German newspaper text from the Frankfurter Rundschau. The corpus was annotated for parts-of-speech and syntax (Skut et al., 1998), and subsequently, also for coreference and entity types (Schiehlen, 2004). It can thus serve to illustrate the problems of multi-layer annotations in linguistic corpora, further, it can be linked to the German DBpedia, and added to Glottolog/Langdoc as a linguistic resource.

Figure 1 illustrates the first sentence[2] in its original representation, a tab-separated text format. (A graphical visualization of this sentence is provided in Chiarcos, this vol., Fig. 1.)

For terminal nodes of the annotation, the first column contains the word, the second a part-of-speech tag, the third morphological annotations, the fourth the label of the edge that connects the terminal with its parent, and the last the ID of the nonterminal parent node. For nonterminal nodes of the annotation, the first column contains the ID, the second the category label, the third is empty, the fourth contains the edge label and the fifth the id of the parent (resp. 0 for the root node).

Traditionally, NEGRA annotations are modified, queried and visualized with specialized tools, e.g., Annotate[3] and Synpathy[4] for syntax annotation, and TIGER-Search (Lezius, 2002) for querying and visualization of syntax annotations. These tools do not, however, allow to annotate and to query over additional layers of annotations, e.g., alignment with parallel corpora, semantic annotations or coreference. For these types of data, further special-purpose tools have been developed, e.g., the Stockholm TreeAligner[5] for aligning parallel corpora annotated with TIGER XML (the XML-based successor of the NEGRA format, König and Lezius, 2000), or SALTO (Burchardt et al., 2006) for the annotation of semantic relations (and also applied to coreference, see Eckart et al., this vol.). These tools are accompanied by other special-purpose formats, e.g., SALSA (Erk and Pado, 2004) for semantic annotations.

As an example for such formats, the coreference annotation of the NEGRA corpus (Schiehlen, 2004) is illustrated in Fig. 3. This is a space-separated text format, where the first column provides the NEGRA sentence id, the second column the token position of a terminal within the the sentence, resp. the id of a NEGRA non-

---

[1] http://www.coli.uni-saarland.de/projects/sfb378/negra-corpus/negra-corpus.html

[2] The examples given in this section are taken from public sample of the NEGRA corpus, the syntax sample is available from http://www.coli.uni-saarland.de/projects/sfb378/negra-corpus/corpus-sample.export, the coreference sample is available from http://www.ims.uni-stuttgart.de/~mike/annotated-negra.txt.

[3] http://www.coli.uni-saarland.de/projects/sfb378/negra-corpus/annotate.html

[4] http://www.lat-mpi.eu/tools/synpathy/

[5] http://kitt.cl.uzh.ch/kitt/treealigner

%% word	tag	morph	edge	parent	gloss
...					
die	ART	Def.Fem.Nom.Sg	NK	507	the
Zukunft	NN	Fem.Nom.Sg.*	NK	507	future
der	ART	Def.Fem.Gen.Sg	NK	502	of
Musik	NN	Fem.Gen.Sg.*	NK	502	music
liegt	VVFIN	3.Sg.Pres.Ind	HD	509	lies
für	APPR	Akk	AC	503	for
viele	PIDAT	*.Akk.Pl	NK	503	many
junge	ADJA	Pos.*.Akk.Pl.St	NK	503	young
Komponisten	NN	Masc.Akk.Pl.*	NK	503	composers
im	APPRART	Dat.Masc	AC	504	in.the
Crossover-Stil	NN	Masc.Dat.Sg.*	NK	504	crossover.style
.	$.	--	--	0	
#502	NP	--	GR	507	
#503	PP	--	MO	509	
#504	PP	--	MO	509	
#507	NP	--	SB	509	
#509	S	--	--	0	
...					

Fig. 1 First sentence of the NEGRA corpus, original NEGRA format with English glosses added. Translation: '... many young composers believe that the future of music lies in a crossover style.'

%% word	tag	morph	edge	parent	gloss
Sie	PPER	3.Pl.*.Nom	SB	504	they
gehen	VVFIN	3.Pl.Pres.Ind	HD	504	enter
gewagte	ADJA	Pos.*.Akk.Pl.St	NK	500	adventurous
Verbindungen	NN	Fem.Akk.Pl.*	NK	500	associations
und	KON	--	CD	502	and
Risiken	NN	Neut.Akk.Pl.*	CJ	502	risks
ein	PTKVZ	--	SVP	504	in
,	$,	--	--	0	

Fig. 2 Second sentence of the NEGRA corpus, original NEGRA format with English glosses added. Translation: 'They experiment with adventurous associations and take risks, ...'

terminal, the third column information about the discourse status of the referent (e.g., R for referring expressions, R1 first mention of a referring expression) and its id (e.g., Komponist for the NEGRA nonterminal corresponding to the phrase *für viele Komponisten* in Fig. 1.

%%sentence	negraid	coref
1	503	%R1=Komponist
2	1	%R=Komponist

Fig. 3 Coreference annotation of the first two sentences of the NEGRA corpus, original format

With comfortable tools for querying and visualization available for syntax only, the question arises how the additional information about coreference can be integrated with other annotation layers, and how queries can be performed on this data.

2 Multi-Layer Corpora as Linked Data
(NEGRA/POWLA Coreference ↦ NEGRA/POWLA Syntax)

In POWLA, NEGRA syntax annotation and coreference annotation can be combined easily as different annotation layers (Chiarcos, this vol.):

```
<!-- syntax "für viele Komponisten" ("for many composers") -->
<powla:Nonterminal rdf:about="s1_503">
   <powla:hasLayer rdf:resource="syntax"/>
   <powla:has_cat>PP</powla:has_cat>
   <powla:hasChild rdf:resource="s1_18"/>
   ...
</powla:Nonterminal>

<!-- syntax "Sie" ("they", sentence 2) -->
<powla:Terminal rdf:about="s2_1">
   <powla:hasLayer rdf:resource="syntax"/>
   <powla:hasString>Sie</powla:hasString>
   <powla:has_pos>PPER</powla:has_pos>
   ...
</powla:Terminal>

<!-- coreference -->
<powla:Relation rdf:about="s2_1_to_s2_530">
   <powla:hasLayer rdf:resource="coref"/>
   <powla:hasSource rdf:resource="s2_1"/>
   <powla:hasTarget rdf:resource="s1_530"/>
</powla:Relation>
```

Using OWL/RDF-based technologies like POWLA, the integration of multiple annotation layers in multi-layer corpora is straight-forward, as previously noticed by Burchardt et al. (2008) for the specific case of syntactic and semantic annotations in the SALSA/TIGER corpus. Formally, different layers can be (but do not have to be) stored in different files and actually at different locations, and can thus be viewed as Linked Data.

Using SPARQL, it is thus possible to query across multiple layers at the same time (which would not have been possible with the original formats and the original tools). For example, we can query for personal pronouns (as defined on the syntactic annotation layer) that take prepositional phrases (defined on syntax, again) as their anaphoric antecedent (coreference layer) using the following query:

```
PREFIX powla:<http://purl.org/powla/powla.owl#>
PREFIX negra:<http://purl.org/powla/negra-sample.owl#>
SELECT ?anaphor
WHERE {
    ?anaphor a powla:Node.
    ?anaphor powla:has_pos "PPER".
    ?relation a powla:Relation.
    ?relation powla:hasSource ?anaphor.
    ?relation powla:hasTarget ?antecedent.
    ?relation powla:hasLayer negra:coref.
    ?antecedent powla:has_cat "PP"
}
```

3 Linking Corpora to Metadata Repositories (POWLA $\mapsto$ Glottolog/Langdoc)

For the linking of linguistic corpora, we take Glottolog language specifications as an example (Nordhoff, this vol.).

POWLA corpora can be linked with glottolog languoid specifications using the Dublin Core language feature.[6] If it is redefined as an owl:ObjectProperty, a POWLA Document, a Terminal or a Nonterminal can be defined as being defined for a particular language:[7]

```
<powla:Document rdf:about="http://purl.org/powla/negra-
                                        sample.owl">
   <dcterms:language
      rdf:resource="http://glottolog.livingsources.org/resource/
                                        languoid/id/10077"/>
</powla:Document>
```

On this basis, we can query for POWLA Documents in German using a simple SPARQL query:

```
PREFIX dcterms: <http://purl.org/dc/terms/>.
SELECT ?doc
WHERE {
   ?doc dcterms:language glottolog:10077
}
```

Alternatively, if we don't know the languoid id, we can query for the label:

---

[6] The description here is somewhat shortened as dcterms:language ranges over dcterms:linguisticSystem, which is not immediately connected to glottolog:languoid. The intermediate steps of the query are not self-evident and will be glossed over here for reasons of space.

[7] Besides Glottolog, other language taxonomies could be applied, e.g., as specified by ISO 639. Glottolog is, however, much more fine-grained than these and captures differentiations that are highly relevant relevant to linguistics, e.g., different dialectal and historical variants of German which are not represented by ISO 639.

```
SELECT ?doc
WHERE {
    ?doc dcterms:language ?languoid.
    ?languoid rdfs:label "German"
}
```

If we know the languoid's ISO 639/3 code,[8] we can query:

```
PREFIX dcterms: <http://purl.org/dc/terms/>.
PREFIX lexvo:   <http://lexvo.org/ontology/>.
SELECT ?doc
WHERE {
  ?doc dcterms:language ?languoid.
  ?languoid lexvo:iso639P3Code "deu"
}
```

In a similar way, other types of metadata can be linked to a linguistic resource, e.g., geographical or historical information.

Glottolog has not been specifically developed for German, instead, it takes a focus on less-resourced languages. However, modeling and querying for corpus resources on, say, the dialects of the indigenous languages of Taiwan documented in the Formosan Languages Archive[9] is analoguous to the treatment of German in this case.

A concrete application of such information for less-resourced languages can be seen, for example, in the context of annotation projection experiments, a flourishing field of Natural Language Processing, where annotated corpora are created on the basis of translated (parallel) text and the assumption that linguistic annotations assigned to word A in the source language are also applicable to the target language word B that A is aligned with. Of course, such experiments benefit from genetic proximity between the language pairs considered, and genetic proximity can be measured by their relative distance in Glottolog/Langdoc classifications or the ASJP tree (Nordhoff, this vol., Sect. 5).[10]

Moreover, Langdoc provides information about the availability of the necessary resources, i.e., translated text (e.g., the Bible), and annotated corpora. By providing this information, and linking to linguistic corpora, Glottolog/Langdoc can thus support the development of NLP resources for languages where such resources are currently not available.

Before starting on an NLP project of Language X, Glottolog/Langdoc allows to check whether sisters (cousins, grandcousins, ...) of language X have already been studied with regard to NLP, and thus jump-start the development.

---

[8] If the languoid does not have an ISO 639 code, information from super- or sublanguoids can be consulted.

[9] http://formosan.sinica.edu.tw

[10] Typological similarity might also be a good indicator for some syntactic questions. This could be retrieved from WALS (Dryer, 2005). The integration of WALS data in the Linguistic Linked Data cloud is currently underway, but not discussed here for reasons of space.

4 Linking Corpora to Terminology Repositories (NEGRA/POWLA ↦ OLiA)

Jsut like linguistic resources can be linked with metadata, annotations can be linked with terminology repositories, so that the semantics of the annotations are represented in an interoperable way. For this purpose, we take the OLiA ontologies as an example (Chiarcos, this vol., Sect. 3). OLiA ontologies provide OWL/DL representations of annotation schemes (OLiA Annotation Model), conventional linguistic terminology (OLiA Reference Model), and that formalize the linking between both.

For the NEGRA corpus, we consider the Annotation Models `stts` for part-of-speech annotation and `tiger-syntax` for syntax annotations according to the TIGER and NEGRA schemes. The Annotation Model defines a taxonomy of linguistic concepts where concrete tags are represented by individuals. For every individual, the property `hasTag`[11] defines its string representation.

For the linking between annotations and OLiA Annotation Models, two strategies are possible: Either we define a property that expresses the linking between annotations and annotation models, or we copy the features of the individual in the Annotation Model to the annotated resource, thereby declaring it an instance of Annotation Model concepts. It should be noted, however, that an annotation may match several individuals, e.g., if tags are composed of multiple independent components, `hasTagMatching` is used in place of `hasTag`. (For example, the `morph` attribute in the NEGRA corpus, every morphological feature – Case, Number, Tense, etc. – is represented by another individual that corresponds to a different substring of the annotation.) In that case, the strategy to copy properties from individuals, rather than to link these individuals, naturally yields an accumulation of all the information expressed in the annotation.

Accordingly, we can define the NEGRA/POWLA entities in terms of OLiA:

```
<powla:Nonterminal rdf:about="s1_503">
  <rdf:type
    rdf:resource="http://purl.org/olia/tiger-syntax.owl#
                                     PrepositionalPhrase"/>
</powla:Nonterminal>

<rdf:Description rdf:about="s1_18">
  <rdf:type
    rdf:resource="http://purl.org/olia/stts.owl#CommonNoun"/>
</rdf:Description>

<rdf:Description rdf:about="s2_1">
  <rdf:type
    rdf:resource="http://purl.org/olia/stts.owl#PersonalPronoun"/>
</rdf:Description>
```

The SPARQL query for PP antecedents of personal pronouns can thus be rephrased as follows:

---

[11] Aside from `hasTag`, there are also properties for the partial matching of strings, e.g., `hasTagContaining` (partial string), `hasTagMatching` (regular expression), etc.

```
PREFIX negra:  <http://purl.org/powla/negra-sample.owl#>.
PREFIX powla:  <http://purl.org/powla/powla.owl#>.
PREFIX stts:   <http://purl.org/olia/stts.owl#>.
PREFIX tiger:  <http://purl.org/olia/tiger-syntax.owl#>.
SELECT ?anaphor
WHERE {
    ?anaphor a stts:PersonalPronoun.
    ?relation a powla:Relation.
    ?relation powla:hasSource ?anaphor.
    ?relation powla:hasTarget ?antecedent.
    ?relation powla:hasLayer negra:coref.
    ?antecedent a tiger:PrepositionalPhrase.
}
```

As concepts of OLiA Annotation Models are linked by `rdfs:subClassOf`
properties to concepts in the OLiA Reference Model, we can infer the corresponding
`rdf:type` properties for the annotated POWLA Nodes:

```
PREFIX negra:  <http://purl.org/powla/negra-sample.owl#>.
PREFIX powla:  <http://purl.org/powla/powla.owl#>.
PREFIX olia:   <http://purl.org/olia/olia.owl#>.
SELECT ?anaphor
WHERE {
    ?anaphor a olia:PersonalPronoun.
    ?relation a powla:Relation.
    ?relation powla:hasSource ?anaphor.
    ?relation powla:hasTarget ?antecedent.
    ?relation powla:hasLayer negra:coref.
    ?antecedent a olia:NounPhrase.
}
```

Despite its name, the OLiA Reference Model does not attempt to establish
a terminological consensus; it only serves as an interface between various Annotation Models and other terminology repositories from which it is derived,
for example, the GOLD ontology (Farrar and Langendoen, 2003; Chiarcos, 2008)
and the ISO TC37/SC4 Data Category Registry (Kemps-Snijders et al., 2008;
Chiarcos, 2010). So, this query can also be formulated in terms of these repositories, e.g., by replacing `olia:PersonalPronoun` with `<http://www.iso-`
`cat.org/datcat/DC-1463>` or `<http://linguistics-ontolo-`
`gy.org/gold/PersonalPronoun>`.

In this way, interoperable corpus queries can be formulated, which can be applied
to corpora with differing annotation schemes.

5 Linking Terminology Repositories to Metadata Repositories (OLiA ↦ Glottolog/Langdoc)

Like linguistic corpora, also other knowledge bases can be augmented with linguistically relevant metadata: Many OLiA Annotation Models, for example the `stts`

and `tiger-syntax` models mentioned above, are specific to a particular corpus, a stage or a particular language, in this case New High German.

This can be expressed, for example, with an axiom that postulates that all instances of the top-level element in an Annotation Model inherit a particular `dcterms:language` property:

```
<owl:Class rdf:about="http://purl.org/olia/stts.owl#Tag">
  <rdfs:subClassOf>
    <owl:Restriction>
      <owl:onProperty>
        <owl:ObjectProperty rdf:about="http://purl.org/dc/terms/
                                        language"/>
      </owl:onProperty>
      <owl:hasValue
        rdf:about="http://glottolog.livingsources.org/resource/
                                        languoid/id/10077"/>
    </owl:Restriction>
  </rdfs:subClassOf>
</owl:Class>
```

With a linking to POWLA data as described above, every POWLA Terminal annotated with `stts` individuals can now be inferred to be New High German.

6 Linking Corpora to Lexical-Semantic Resources (NEGRA/POWLA ↦ DBpedia)

As described by Hellmann et al. (this vol.), textual data can be automatically enriched with entity links, e.g., to the (German instantiation of the) German DBpedia, e.g., using a NIF-based NLP pipeline. The development of the NLP Interchange Format is synchronized with the development of POWLA. Although both are optimized for different purposes – POWLA is developed to represent annotated corpora with a high degree of genericity, whereas NIF is a compact and NLP-specific format[12] –, they are designed to be mappable. This means that NIF annotations can be converted to POWLA representations, and then, for example, combined with other annotation layers.

---

[12] One difference is the representation of labeled relations between two entities: In NIF, these are represented as properties, with the ID reflecting the annotation attached to the relation. The NIF modeling requires only a single triple per relation. In POWLA, however, a labeled relation is an individual that is linked by properties to its source and target, and that is assigned its annotation by another property. The POWLA modeling requires at least four triples per relation. Unlike NIF, however, this modeling allows to attach complex annotations to relations, e.g., a direct linking to the OLiA concept hierarchy.
Another difference is that NIF lacks any formalization of corpus structure and annotation layers. More important is that, at the moment, NIF is capable to represent morphosyntactic and syntactic annotations only, the representation of more complex forms of annotation, e.g., alignment in a parallel corpus, has not been addressed so far.
The NIF representation is thus more compact, but the POWLA representation is more precise and more expressive.

As a result, the POWLA individual `negra:s1_18` (*Komponisten* 'composers'), for example, can be annotated with the corresponding DBpedia concept:

```
<powla:Terminal rdf:about="s1_18">
  <powla:hasString>Komponisten</powla:hasString>
  <scms:means
    rdf:resource="http://de.dbpedia.org/resource/Komponist"/>
  ...
</powla:Terminal>
```

As generated by DBpedia spotlight, this information is attached to POWLA `Terminals` (words), and it can easily be projected further to the corresponding phrases:

```
PREFIX powla: <http://purl.org/powla/powla.owl#>.
PREFIX olia:  <http://purl.org/olia/olia.owl#>.
PREFIX scms:<http://ns.aksw.org/scms/>.
CONSTRUCT { ?np scms:means ?semClass }
SELECT {
    ?term a olia:Noun.
    ?term scms:means ?semClass.
    ?np powla:hasChild ?term.
    ?np a olia:NounHeadedPhrase
}
```

For pronouns with `NounPhrase` or `PrepositionalPhrase` antecedents[13] this information can be used to assign them a semantic class:

```
PREFIX negra: <http://purl.org/powla/negra-sample.owl#>.
PREFIX powla: <http://purl.org/powla/powla.owl#>.
PREFIX olia:  <http://purl.org/olia/olia.owl#>.
PREFIX scms:  <http://ns.aksw.org/scms/>.
CONSTRUCT { ?np scms:means ?semClass }
WHERE {
    ?relation a powla:Relation.
    ?relation powla:hasLayer negra:coref.
    ?relation powla:hasSource ?pronoun.
    ?pronoun a olia:Pronoun.
    ?relation powla:hasTarget ?antecedent.
    ?antecedent scms:means ?semClass.
}
```

By combining information from POWLA, OLiA and DBpedia, we can thus achieve richer semantic annotations for linguistic corpora, that can then be used, for example, to develop NLP applications on this basis, e.g., an anaphor resolution system that takes DBpedia categories into account (Bryl et al., 2010).

---

[13]   `NounHeadedPhrase` is a generalization over `NounPhrase` and `PrepositionalPhrase` that was introduced to account for annotation schemes where both are not properly distinguished.

7 Enriching Lexical-Semantic Resources with Linguistic Information (DBpedia (↦ POWLA) ↦ OLiA)

Unlike classical lexical-semantic resources, DBpedia offers almost no information about the linguistic realization of the entities it contains. Using corpora with entity links and syntactic annotation, however, this information can be easily obtained. The following SPARQL query identifies possible syntactic realizations for concepts of the German DBpedia:

```
PREFIX powla: <http://purl.org/powla/powla.owl#>.
PREFIX olia:  <http://purl.org/olia/olia.owl#>.
PREFIX scms:  <http://ns.aksw.org/scms/>.
CONSTRUCT { ?semClass <#realizedAs> ?syntClass }
WHERE {
  ?x a powla:Node.
  ?x scms:means ?semClass.
  ?x a ?syntClass
  FILTER(regex(str(?syntClass),"http://purl.org/olia/olia.owl#")).
  ?syntClass rdfs:subClassOf olia:MorphosyntacticCategory.
}
```

With the newly generated triples added to the DBpedia, this information about possible grammatical realizations of an entity can be used, for example, to enhance entity-linking algorithms.

8 Enriching Lexical-Semantic Resources with Metadata (DBpedia (↦ POWLA) ↦ Glottolog)

Similarly, lexical-semantic resources can be enriched with metadata. With fine-grained languoid definitions as provided by Glottolog, and corpora representing different historical stages of a language, for example, the historical development of terms can be extrapolated (when and where was a term recorded).

The following SPARQL query assigns a concept a dcterms:language property for every languoid for which it is found in a particular corpus:

```
PREFIX powla:   <http://purl.org/powla/powla.owl#>.
PREFIX dcterms: <http://purl.org/dc/terms/>.
PREFIX scms:    <http://ns.aksw.org/scms/>.
CONSTRUCT {?semClass dcterms:language ?language}
WHERE {
    ?word a powla:Node.
    ?word dcterms:language ?language.
    ?word scms:means ?semClass.
}
```

This query presupposes that entity linking has been performed on the corpus before. On a historical or dialectal corpus, entity linking is possible only if its spelling follows the same conventions as modern standard orthography for the respective language. However, dialectal and historical corpora often include lemmas in modern languages, and on these 'hyperlemmas' (Dipper et al., 2004), a standard entity linking routine can be applied.

9 Enriching Metadata Repositories with Linguistic Features (Glottolog ↦ OLiA)

Finally, one may consider also to enrich metadata repositories with linguistic features, e.g., to record which languoid makes use of which linguistic categories and features.

On the basis of the resources described before, this can be extrapolated from annotations in a languoid-annotated corpus.[14] The following query retrieves all syntactic categories that are used for a particular Glottolog languoid (given a set of corpora to which this query is applied):

```
PREFIX dcterms:   <http://purl.org/dc/terms/>.
PREFIX powla:     <http://purl.org/powla/powla.owl#>.
PREFIX olia:      <http://purl.org/olia/olia.owl#>.
PREFIX rdfs:      <http://www.w3.org/2000/01/rdf-schema#>.
CONSTRUCT { ?languoid <#uses> ?syntacticCategory }
WHERE {
   ?node dcterms:language ?languoid
   FILTER(regex(str(?languoid),"http://glottolog.livingsources.
                                 org/resource/languoid/id/.*")).
   ?node a powla:Node.
   ?node a ?syntacticCategory
   FILTER(regex(str(?syntacticCategory),
              "http://purl.org/olia/olia.owl#.*")).
   ?syntacticCategory rdfs:subClassOf olia:SyntacticCategory.
}
```

On this basis, then, one may study to what extent genealogical relationships correspond to certain syntactic features (as far as reflected in the underlying resources). For instance, one might build a reasoner which asserts the existence of a grammatical category to a `glottolog:superlanguoid` if all its sublanguoids happen to have this particular property. For instance, the category 'Preposition' is found in

---

[14] It should be noted that this approach is *approximative* only, because it considers only information expressed in annotations. If is possible that the underlying schemes make a number of simplifying assumptions, e.g., not to distinguish two functionally different categories that appear superficially and that cannot be unambiguously distinguished by NLP tools or human annotators. Greater precision could probably be achieved if such queries are applied to language-annotated lexicons that make use of a standard vocabulary to represent detailed grammatical information, as created, for example, in the context of the LEGO project (Poornima and Good, 2010) whose lexicons are linked to the GOLD ontology (Farrar and Langendoen, 2003). The queries necessary for this purpose would be, however, almost identical.

corpora of German, Dutch, English, and all other Germanic languages. Such a category can therefore be posited on the family level. Postpositions on the other hand are only found in a subset of the Germanic languages and thus do not 'climb up the tree' as high as their prenominal brethren.

If knowledge bases are provided that provide other metrics of language relatedness (e.g., ASJP, Nordhoff, this vol., Sect. 5), it can be tested whether these metrics correspond to the occurrence of similar grammatical features. The Linked Data approach furthermore allows to map nodes of different trees to each other. Computation of consensus trees from trees based on different datasets is another possibility.

10 Outlook

We illustrated how a Linguistic Linked Open Data cloud can be created, and what possible gains of information would be possible. The resources described in this part and their possible linking are summarized in Fig. 4.

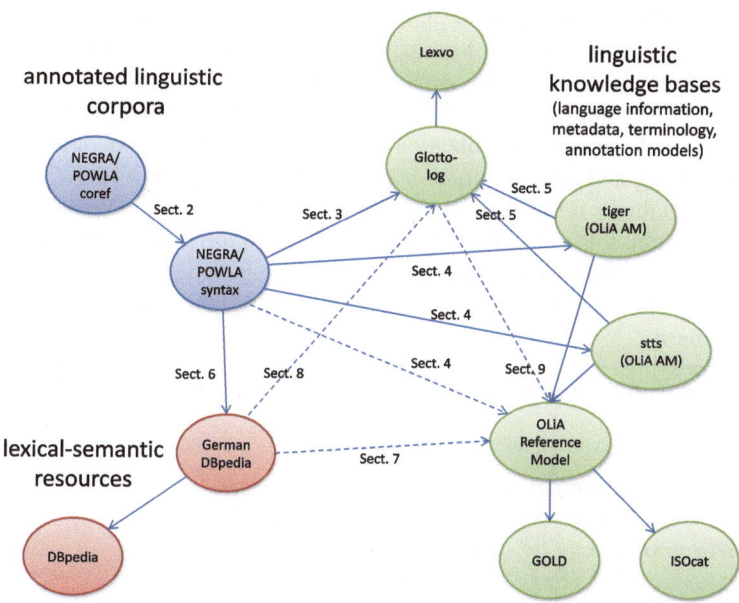

Fig. 4 Linguistic resources and possible links between them as described in this contribution.

It should be noted that the resources considered here only serve illustrative purposes, and they are chosen such that they represent major types of linguistic resources: lexical-semantic resources, linguistic corpora and repositories of metadata and linguistic terminology. Various members of the OWLG are engaged in related efforts, and we expect that these converge in the creation of a Linguistic Linked Open Data cloud as described here. As of December 2011, a number of OWLG members have expressed their willingness to provide data for such a Linguistic Linked Open Data cloud, aside from the resources shown in Fig. 4 this includes RDF formalizations of WordNet and Wiktionary (e.g., those described by McCrae et al., this vol.) and other lexical-semantic resources, typological metadata repositories (e.g., those described by Moran, this vol.), additional corpora, and multi-lingual word lists.

These novel resources are complemented by linguistic resources already present in the Linked Open Data cloud,[15] e.g., meta data repositories such as Lexvo,[16] lingvoj,[17] GeoNames,[18] Project Gutenberg,[19] the OpenLibrary;[20] and lexical-semantic resources such as Cornetto,[21] Freebase,[22] OpenCyc,[23] the Open Data Thesaurus,[24] YAGO,[25] and WordNet.[26] Additionally, the OWLG has compiled an extensive list of resources that represent candidates to be included in the LLOD and that are available under open licenses, but that have not yet been converted to RDF.

It is our hope that the workshop on Linked Data in Linguistics (LDL-2012) and this volume contribute to the on-going formation of an interdisciplinary community actively working towards the application of the Linked Open Data paradigm to all forms of linguistic resources. In the last few years, interested researchers in different sub-communities have begun to organize themselves. One example is the Open Linguistics Working Group,[27] founded in 2009, who organized LDL-2012, another one is the W3C Ontology-Lexica Community Group, founded in 2011.[28] At the same time, established standardization initiatives from the fields of Natural Language Processing and computational lexicography, e.g., the ISO TC37/SC4, adopt ideas and formalisms developed in the context of the Semantic Web (e.g., Windhouwer and Wright, this vol.; Pareja-Lora, this vol.). In the fields of language documentation, typology and in the humanities in general (e.g., Bouda and Cysouw, this vol.; Declerck et al., this vol.; Schalley, this vol.), Linked Data approaches seem to be gaining popularity.

---

[15] http://richard.cyganiak.de/2007/10/lod

[16] http://www.lexvo.org

[17] http://www.lingvoj.org

[18] http://www.geonames.org/ontology

[19] http://www4.wiwiss.fu-berlin.de/gutendata

[20] http://openlibrary.org

[21] http://www2.let.vu.nl/oz/cltl/cornetto

[22] http://freebase.com

[23] http://sw.opencyc.org

[24] http://vocabulary.semantic-web.at/PoolParty/wiki/OpenData

[25] http://mpii.de/yago

[26] http://semanticweb.cs.vu.nl/lod/wn30, http://www.w3.org/TR/word net-rdf, http://wordnet.rkbexplorer.com

[27] linguistics.okfn.org

[28] http://www.w3.org/community/ontolex

In this chapter, we have discussed some first nodes of a Linguistic Linked Data Cloud. We have also discussed how links between these nodes can be established. The wide range of topics covered in this volume as well as the commitment shown by scholars from very different subdisciplines of linguistics to render their data interoperable make us very optimistic that this network will quickly grow and that the coverage of the LLD cloud as well as its density will significantly increase in the very near future.

References

Bouda P, Cysouw M (this vol.) Treating dictionaries as a Linked-Data corpus. pp 15–23

Bryl V, Giuliano C, Serafini L, Tymoshenko K (2010) Supporting natural language processing with background knowledge: coreference resolution case. The Semantic Web–ISWC 2010, pp 80–95

Burchardt A, Erk K, Frank A, Kowalski A, Pado S (2006) SALTO: A versatile multi-level annotation tool. In: Proc. LREC-2006, Genoa, Italy

Burchardt A, Padó S, Spohr D, Frank A, Heid U (2008) Formalising Multi-layer Corpora in OWL/DL – Lexicon Modelling, Querying and Consistency Control. In: Proceedings of the 3rd International Joint Conference on NLP (IJCNLP 2008), Hyderabad

Chiarcos C (2008) An ontology of linguistic annotations. LDV Forum 23(1):1–16

Chiarcos C (2010) Grounding an ontology of linguistic annotations in the Data Category Registry. In: LREC 2010 Workshop on Language Resource and Language Technology Standards (LT&LTS), Valetta, Malta, pp 37–40

Chiarcos C (this vol.) Interoperability of corpora and annotations. pp 161–179

Declerck T, Lendvai P, Mörth K, Budin G, Váradi T (this vol.) Towards Linked Language Data for Digital Humanities. pp 109–116

Dipper S, Faulstich L, Leser U, Lüdeling A (2004) Challenges in modelling a richly annotated diachronic corpus of German. In: Workshop on XML-based richly annotated corpora, Lisbon, Portugal, pp 21–29.

Dryer MS (2005) Genealogical language list. In: Comrie B, Dryer MS, Gil D, Haspelmath M (eds) World Atlas of Language Structures. Oxford University Press, pp 584–644

Eckart K, Riester A, Schweitzer K (this vol.) A discourse information radio news database for linguistic analysis. pp 65–75

Erk K, Pado S (2004) A powerful and versatile XML format for representing role-semantic annotation. In: Proc. Fourth International Conference on Language Resources and Evaluation (LREC), Lisbon, Portugal

Farrar S, Langendoen DT (2003) A Linguistic Ontology for the Semantic Web. GLOT International 7:97–100

Hellmann S, Stadler C, Lehmann J (this vol.) The German DBpedia: A sense repository for linking entities. pp 181–189

Kemps-Snijders M, Windhouwer M, Wittenburg P, Wright S (2008) ISOcat: Corralling data categories in the wild. In: Proc. LREC 2008, Marrakech, Morocco

König E, Lezius W (2000) A description language for syntactically annotated corpora. In: Proc. 18th International Conference on Computational Linguistics (COLING 2000), Saarbrücken, Germany, pp 1056–1060

Lezius W (2002) TIGERSearch. Ein Suchwerkzeug für Baumbanken. In: Proceedings of the 6. Konferenz zur Verarbeitung natürlicher Sprache (6th Conference on Natural Language Processing, KONVENS 2002), Saarbrücken, Germany

McCrae J, Montiel-Ponsoda E, Cimiano P (this vol.) Integrating WordNet and Wiktionary with *lemon*. pp 25–34

Moran S (this vol.) Using Linked Data to create a typological knowledge base. pp 129–138

Nordhoff S (this vol.) Linked Data for linguistic diversity research: Glottolog/Langdoc and ASJP. pp 191–200

Pareja-Lora A (this vol.) OntoLingAnnot's ontologies: Facilitating interoperable linguistic annotations (up to the pragmatic level). pp 117–127

Poornima S, Good J (2010) Modeling and encoding traditional wordlists for machine applications. In: Proceedings of the 2010 Workshop on NLP and Linguistics: Finding the Common Ground, Association for Computational Linguistics, pp 1–9

Schalley AC (this vol.) TYTO – A collaborative research tool for linked linguistic data. pp 139–149

Schiehlen M (2004) Optimizing algorithms for pronoun resolution. In: Proc. 20th International Conference on Computational Linguistics (COLING), Geneva, pp 515–521

Skut W, Brants T, Krenn B, Uszkoreit H (1998) A linguistically interpreted corpus of German newspaper text. In: Proc. ESSLLI Workshop on Recent Advances in Corpus Annotation, Saarbrücken, Germany

Windhouwer M, Wright SE (this vol.) Linking to linguistic data categories in ISOcat. pp 99–107